PREFACE

Currently there is a vast amount of literature on nonlinear programming in finite dimensions. The publications deal with convex analysis and several aspects of optimization. On the conditions of optimality they deal mainly with generalizations of known results to more general problems and also with less restrictive assumptions. There are also more general results dealing with duality. There are yet other important publications dealing with algorithmic development and their applications. This book is intended for researchers in nonlinear programming, and deals mainly with convex analysis, optimality conditions and duality in nonlinear programming. It consolidates the classic results in this area and some of the recent results.

The book has been divided into two parts. The first part gives a very comprehensive background material. Assuming a background of matrix algebra and a senior level course in Analysis, the first part on convex analysis is self-contained, and develops some important results needed for subsequent chapters. The second part deals with optimality conditions and duality. The results are developed using extensively the properties of cones discussed in the first part. This has facilitated derivations of optimality conditions for equality and inequality constrained problems. Further, minimum-principle type conditions are derived under less restrictive assumptions. We also discuss constraint qualifications and treat some of the more general duality theory in nonlinear programming.

Atlanta, Georgi
December 1975

1.S. Bazaraa
C.M. Shetty

Lecture Notes in Economics and Mathematical Systems

Managing Editors: M. Beckmann and H. P. Künzi

Mathematical Programming

122

M. S. Bazaraa
C. M. Shetty

Foundations of Optimization

Springer-Verlag
Berlin · Heidelberg · New York 1976

Authors
M. S. Bazaraa
C. M. Shetty
Georgia Institute of Technology
School of Industrial and Systems Engineering
Atlanta, GA 30332/USA

Library of Congress Cataloging in Publication Data

Bazaraa, M S 1943-
Foundations of optimization.

(Lecture notes in economics and mathematical systems ; 122)
Bibliography: p.
Includes index.
1. Mathematical optimization. 2. Nonlinear programming. 3. Duality theory (Mathematics)
I. Shetty, C. M., 1929- joint author.
II. Title. III. Series.
QA402.5.B39 519.7'6 76-6574

AMS subject Classifications (1970): 90C25, 90C30, 52A20, 54C30, 54-01

ISBN 3-540-07680-8 Springer-Verlag Berlin · Heidelberg · New York
ISBN 0-387-07680-8 Springer-Verlag New York · Heidelberg · Berlin

Printed in Germany

Printing and binding: Beltz Offsetdruck, Hemsbach/Bergstr.
2142/3140-543210

TABLE OF CONTENTS

Part I: Convex Analysis

Part II: Optimality Conditions and Duality

CHAPTER 1

LINEAR SUBSPACES AND AFFINE MANIFOLDS

The first three chapters of this book discuss the concept of linear subspaces and some of its important subsets -- namely, affine manifolds, convex cones and sets. The notion of convexity plays a dominant role in nonlinear programming and is explored in depth in these chapters. Chapter 4 deals with convex and convex-like functions. As we will see later, certain convex sets can be associated with each of these functions. The important results of these chapters are used later to develop optimality conditions for nonlinear programs. The chapters also do contain several other related results which have been used elsewhere in the study of nonlinear programs or, in the opinion of the authors, are likely to be useful in advanced work in this area.

1.1 Linear Subspaces and Orthogonal Complements

1.1.1 **Definition.** Let $x_1,x_2,\ldots,x_k$ be points in E_n. Then $x = \sum_{i=1}^{k} \lambda_i x_i$ where $\lambda_i \in E_1$, $i = 1,2,\ldots,k$ is said to be a linear combination of $x_1,x_2,\ldots,x_k$. Furthermore if $\lambda_i > 0$ ($\lambda_i \geq 0$) for each i, then x is said to be a positive (nonnegative) linear combination of $x_1,x_2,\ldots,x_k$. x is said to be semipositive combination of $x_1,x_2,\ldots,x_k$ if it is a nonzero nonnegative combination of them. If in addition $\sum_{i=1}^{k} \lambda_i = 1$ then x is called a convex combination of $x_1,x_2,\ldots,x_k$.

The above heirarchy will be found useful in discussing linear subspaces, affine manifolds, convex cones, and convex sets which are discussed in the first three chapters.

1.1.2 Definition. Let L be a set in E_n. L is called a subspace (or linear subspace) if $x_1,x_2 \in L$ imply that $\lambda_1 x_1 + \lambda_2 x_2 \in L$ for each $\lambda_1,\lambda_2 \in E_1$.

Put differently, a set in E_n is a subspace if for any two points in the set, all linear combinations of these points belong to the set. Of course, without any additional generality, we can state Definition 1.1.2 above in terms of any finite number of points. The reader may verify this statement by a simple induction argument.

It is obvious that the origin is always a member of any nonempty subspace by letting all λ's to be zero. Trivial examples of subspaces in E_n are the empty set $\emptyset$, the origin, and E_n itself. Examples of subspaces in the plane E_2 are lines through the origin.

Given two subspaces L_1 and L_2 we may be interested in finding the largest possible subspace L contained in both L_1 and L_2 as well as the smallest possible subspace L' which contains both L_1 and L_2. It can be immediately checked that $L_1 \cap L_2$ is a subspace and since it is the largest set contained in both L_1 and L_2, it follows that $L = L_1 \cap L_2$. On the other hand if there is a subspace which contains both L_1 and L_2, then by Definition 1.1.2 it contains their sum $L_1 + L_2$. Since the latter is a subspace then $L' = L_1 + L_2$. From this discussion the following is obvious.

1.1.3 <u>Lemma</u>. Let L_1 and L_2 be two subspaces in E_n. Then the largest possible subspace contained in both L_1 and L_2 is $L_1 \cap L_2$ and the smallest possible subspace containing both L_1 and L_2 is $L_1 + L_2$.

The reader may be tempted to believe that since $L_1 \cup L_2$ is the <u>smallest set</u> containing both L_1 and L_2 then $L_1 + L_2$ in the above remark should be replaced by $L_1 \cup L_2$. However this cannot be done because $L_1 \cup L_2$ is not necessarily a subspace even if L_1 and L_2 are. For example, if L_1 and L_2 are distinct lines through the origin in E_2 then their union is obviously not a subspace.

If we have two subspaces L_1 and L_2 then their sum L is given by $L_1 + L_2$. If in addition we have $L_1 \cap L_2 = \{0\}$ then L is called the <u>direct sum</u> of L_1 and L_2 and is denoted by $L = L_1 \oplus L_2$. This notion will be useful when we decompose E_n into the direct sum of two orthogonal subspaces.

Given an arbitrary set S in E_n one can generate a subspace which is spanned by S. This subspace is given by the following definition.

1.1.4 <u>Definition</u>. Let S be an arbitrary set in E_n. The <u>subspace spanned (or generated)</u> by S and denoted by L(S) is the set of all linear combinations of points in S, i.e., $L(S) = \{x : x = \sum_{i=1}^{k} \lambda_i x_i,\ x_i \in S,\ \lambda_i \in E_1,\ k \geq 1\}$. A set S is said to <u>span (generate)</u> a subspace L if $L = L(S)$.

The reader can easily verify that L(S) is indeed a subspace according to Definition 1.1.2. Some examples of linear subspaces generated by sets are given below.

(i) $S = \{(x,y): 0 \le x \le 1, y = 0\}$

$L(S) = \{(x,y): x \in E_1, y = 0\}$

(ii) $S = \{(x,y): 0 \le x \le 1, y = x + 1\}$

$L(S) = E_2$

(iii) $S = \{(1,0), (1,1)\}$

$L(S) = E_2$

(iv) By convention if $S = \emptyset$ then $L(S) = \emptyset$.

The reader may note that the set S of the above definition may or may not consist of a finite number of points. We will show in Section 1.2 that any linear subspace can be generated by a finite number of points, i.e., given a subspace L we can find a finite set S with $L(S) = L$.

Actually given a set S in E_n, the subspace L(S) spanned by S has the following interesting property. L(S) is the smallest possible linear subspace which contains S. This fact is obvious since any subspace which contains S will also contain all linear combinations of S, namely L(S). Hence the following lemma is immediate.

1.1.5 Lemma. L(S) is the smallest subspace containing S.

Corollary.

(i) L(S) is the intersection of all subspaces containing S.

(ii) S is a subspace if and only if $L(S) = S$.

The following lemma that can be easily verified gives the relationship between the subspaces spanned by two subsets, and also the relationship between the subspace spanned by the intersection of two arbitrary sets and the intersection of the subspaces spanned by the individual sets.

1.1.6 Lemma.

(i) If S_1 and S_2 are arbitrary sets in E_n with $S_1 \subset S_2$ then $L(S_1) \subset L(S_2)$.

(ii) $L(S_1 \cap S_2) \subset L(S_1) \cap L(S_2)$.

It may be noted that in the second part of the above lemma, the reverse inclusion does not hold in general. For example let $S_1 = \{(x,y): x,y \ge 0\}$ and

$S_2 = \{(x,y): x,y \le 0\}$. Therefore $L(S_1) = L(S_2) = E_2$ whereas $S_1 \cap S_2 = \{(0,0)\}$ and hence $L(S_1 \cap S_2) = \{(0,0)\}$.

Closely associated with a subspace L is another subspace called the orthogonal complement of L and denoted by $L^{\perp}$. The following is a definition of $L^{\perp}$.

1.1.7 Definition. Let L be a subspace in E_n. The orthogonal complement of L, denoted by $L^{\perp}$, is the set of vectors which are orthogonal (perpendicular) to each point in L, i.e., $L^{\perp} = \{y: \langle x,y\rangle = 0 \text{ for each } x \in L\}$. By convention if L is the empty set then $L^{\perp}$ is E_n.

It is obvious that L and $L^{\perp}$ have no points in common other than the origin because if we let $x \in L \cap L^{\perp}$ then $0 = \langle x,x\rangle = \| x \|^2$ which implies that $x = 0$.

1.2. Linear Independence and Dimensionality

1.2.1 Definition. The vectors $a_1,a_2,\ldots,a_k$ in E_n are said to be linearly independent if $\sum_{i=1}^{k} \alpha_i a_i = 0$ implies that $\alpha_1 = \alpha_2, \ldots = \alpha_k = 0$. $a_1,a_2,\ldots,a_k$ are called linearly dependent if they are not linearly independent.

From the above definition it is clear that $a_1,a_2,\ldots,a_k$ are linearly dependent if and only if some a_i can be represented as a linear combination of the other vectors. Letting $A_i = \{a_j : j \neq i\}$ then the vectors $a_1,a_2,\ldots,a_k$ are linearly dependent if and only if for some i we have $a_i \in L(A_i)$, the linear subspace spanned by A_i. The set is linearly independent if and only if for each i we have $a_i \notin L(A_i)$.

From the above definition, it is clear that the zero vector itself forms a linearly dependent set. Hence any set of vectors containing the zero vector is linearly dependent.

We will now consider the notion of bases of a subspace and the notion of a dimension of a subspace.

1.2.2 Definition. Let L be a linear subspace and $a_1,a_2,\ldots,a_k$ be linearly independent vectors in L. If $L(a_1,a_2,\ldots,a_k)$, the subspace spanned by $a_1,a_2,\ldots,a_k$, is equal to L then the set of vectors $a_1,a_2,\ldots,a_k$ is said to form a basis of L. The dimension of L, denoted by dim L is then equal to k. By convention we will let $\dim \emptyset = -1$, and $\dim \{0\} = 0$.

It is appropriate to ask whether or not the notion of dimensionality is well defined. In other words, is it possible for a subspace to have two bases with different number of elements? If this were the case then the notion of dimension of a subspace would not be well defined. However, this can never be the case as given by the corollary to Theorem 1.2.3 below.

1.2.3 Theorem. Let $B = \{b_1, b_2, \ldots, b_r\}$ be a basis of a linear subspace L. Then any set of $r + 1$ vectors in L are dependent.

Proof: Suppose that one can find vectors $a_1, a_2, \ldots, a_{r+1}$ independent vectors in L. Since B spans L then a_1 can be represented as a linear combination of $b_1, b_2, \ldots, b_r$, i.e., $a_1 = \sum_{i=1}^{r} \alpha_{1i} b_i$. Not all the α_{1i}'s are zero and without loss of generality assume that $\alpha_{1i} \neq 0$. Therefore $b_1 = \frac{1}{\alpha_{11}} a_1 - \sum_{i=2}^{r} \frac{\alpha_{1i}}{\alpha_{11}} b_i$. This shows that $a_1, b_2, \ldots, b_r$ span L. Similarly b_2 can be represented as a linear combination of $a_1, a_2, b_3, \ldots, b_r$ and hence they span L. Continuing in this fashion it follows that $a_1, a_2, \ldots, a_r$ span L. This means that a_{r+1} can be written as a linear combination of $a_1, a_2, \ldots, a_r$ which violates the independence assumption.

Corollary. Any two bases of a subspace must contain the same number of elements.

Proof: Let $B = \{b_1, b_2, \ldots, b_r\}$ and $A = \{a_1, a_2, \ldots, a_s\}$ be two bases. Suppose that $s \geq r$. By Theorem 1.2.3, s cannot be strictly greater than r and hence $s = r$.

From the above the reader may note that if L_1 and L_2 are linear subspaces with $L_1 \subset L_2$ then $\dim L_1 \leq \dim L_2$. Also it is worthwhile mentioning that given a basis B of a linear subspace L, then any point in L can be uniquely written as a linear combination of elements. Uniqueness of the representation can be easily established as follows. Let $x = \sum_{i=1}^{k} \alpha_i b_i$ and also $x = \sum_{i=1}^{k} \beta_i b_i$ where $b_1, b_2, \ldots, b_k$ is a basis of L. Hence we get $\sum_{i=1}^{k} (\alpha_i - \beta_i) b_i = 0$ and since $b_1, b_2, \ldots, b_k$ are independent then $\alpha_i = \beta_i$.

The following Theorem shows that any subspace in E_n is spanned by a finite number of points in the subspace. Needless to say that not all sets in E_n has this property, for example any circle in E_2 cannot be generated by a finite number of points from the circle.

1.2.4 Theorem. Let L be a nontrivial subspace in E_n (i.e., neither empty nor the subspace with only the zero vector). Then one can find linearly independent points $a_1, a_2, \ldots, a_r$ in L which span it with $r \leq n$, and hence dim L = $r \leq n$.

Proof: Pick any nonzero $a_1 \in L$. If $L(a_1) = L$ then the result follows. If $L(a_1) \neq L$ then pick a point a_2 in L but not in $L(a_1)$. It is clear that a_1 and a_2 are linearly independent. Now pick a point a_3 in L but not in $L(a_1, a_2)$ if they are not equal. Continuing in this fashion we generate a set of independent vectors. This process will step in at most n iterations since there can be no n + 1 independent vectors in E_n. This shows that we can find $a_1, a_2, \ldots, a_r$ in L with $r \leq n$ such that $L(a_1, a_2, \ldots, a_r) = L$ and the proof is complete.

From the above Theorem it follows that any subspace L in E_n can be written as the set $\{Ax: x \in E_r\}$ where A is an $n \times r$ matrix formed by a basis of L with $r \leq n$. Note that the columns of the matrix A, namely $a_1, a_2, \ldots, a_r$ are independent.

Based upon this characterization of a subspace we can show an important characteristic of a subspace, namely each subspace in E_n is a closed set. Now consider the following definition of the closure of a set.

1.2.5 Definition. Let S be an arbitrary set in E_n. The closure of S denoted by $c\ell\, S$ is the set defined by $c\ell\, S = \{x: N_\epsilon(x) \cap S \neq \emptyset \text{ for each } \epsilon > 0\}$. S is said to be closed if $S = c\ell\, S$.

From this definition it follows that $x \in c\ell\, S$ if and only if there is a sequence x_n is S which converges to x. This is obvious by using the above definition and using a sequence $\{\epsilon_k\}$ converging to zero and for each $\epsilon_k > 0$, picking a point $x_k \in N_{\epsilon_k}(x) \cap S$. From this it is also obvious that a set is closed if and only if any convergent sequence in S has its limit point in S.

Now we will show that a subspace in E_n is indeed closed. We will use this fact in developing the projection theorem of next section.

1.2.6 Lemma. Let L be a linear subspace in E_n. Then L is closed.

Proof: As a result of Theorem 1.2.4, L can be represented by $L = \{Ax: x \in E_r\}$ where A is an $n \times r$ matrix with independent columns. To show that L is closed we need to show that given any convergent sequence $\{y_k\}$ in L has its limit y in L. If we let $y_k = Ax_k$, then $A^t y_k = A^t A x_k$ and hence $x_k = (A^t A)^{-1} A^t y_k = P y_k$ where

$P = (A^tA)^{-1} A^t$ defines a continuous linear operator. Hence $\| x_k \| = \| P y_k \| \leq \|P\| \cdot \|y_k\|$ where $\| P \|$ is the norm of the matrix P defined by $\sup_{\|z\| \leq 1} \| Pz\|$. Now let $y_k \to y$. Then, since $\{x_k\}$ is bounded, it has a convergent subsequence x_{k_j} with a limit x. That is, we have $y_{k_j} = A x_{k_j}$ with $y_{k_j} \to y$ and $x_{k_j} \to x$. This implies $y = A x$, i.e. $y \in L$. Hence L is closed.

1.3 Projection Theorem

Here we will show that given a point $x \in E_n$ and a nonempty subspace L, there is a unique point x_o in L with minimum distance from x. The point x_o is characterized by $x-x_o \in L^{\perp}$ and hence we can develop an orthogonal projection of x onto L.

Before giving the main Theorem, the following result, usually referred to as the parallelogram law, is needed. The law says that the sum of the square of the lengths of the diagonals of a parallelogram is equal to the sum of the square of the lengths of its sides. See Figure 1.1

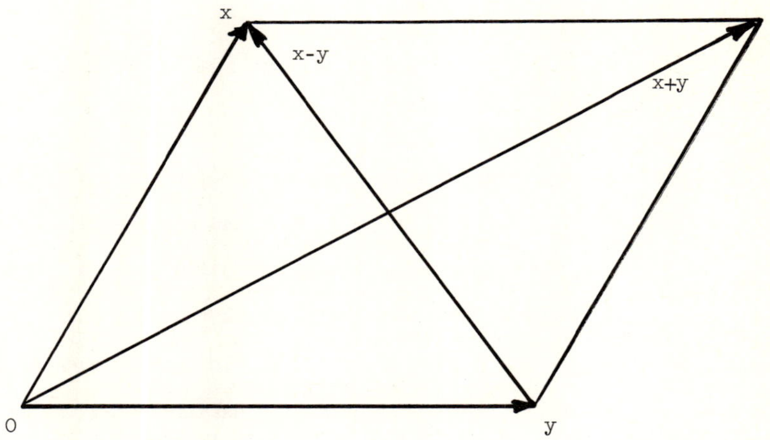

Figure 1.1.

The Parallelogram Law

1.3.1 Lemma (Parallelogram Law). Let $x,y \in E_n$. Then

$$\| x + y\|^2 + \|x - y\|^2 = 2\| x\|^2 + 2\|y\|^2 .$$

Proof: Noting that $\|x + y\|^2 = \langle x + y, x + y\rangle$ and $\|x - y\|^2 = \langle x - y, x - y\rangle$ we get $\|x + y\|^2 + \|x - y\|^2 = \|y\|^2 + 2\langle x,y\rangle + \|x\|^2 + \|y\|^2 - 2\langle x,y\rangle = 2\|x\|^2 + 2\|y\|^2$.

The following definition of the distance between a point and a set is convenient.

1.3.2 Definition. Let $x \in E_n$ and S be a subset of E_n. The distance between x and S, denoted by $d(x,S)$, is given by $d(x,S) = \inf \{\|x - y\| : y \in S\}$. By convention if S is empty then $d(x,S) = \infty$.

1.3.3 Theorem (Projection Theorem). Let $x \in E_n$ and L be a subspace of E_n. Then there exists a unique point $x_o \in L$ with $\|x - x_o\| = d(x,L) = \min \{\|x - y\| : y \in L\}$. Furthermore x_o is such a unique minimizing point if and only if $x - x_o \in L^\perp$.

Proof: Suppose that $x \notin L$ because otherwise the result is trivial. Let $\delta = d(x,L)$ and let $\{x_k\}$ be a sequence in L such that $\|x - x_k\| \to \delta$ (such a sequence exists by definition of δ). We will show that $\{x_k\}$ is a Cauchy sequence and hence has a limit x_o. Since L is closed by Lemma 1.2.6 then $x_o \in L$. By the parallelogram law we get:

$$\|x_n - x_m\|^2 = 2\|x_n - x\|^2 + 2\|x_m - x\|^2 - 4\|x - \frac{x_n + x_m}{2}\|^2 .$$

But $\frac{x_n + x_m}{2} \in L$ since the latter is a subspace and so $\|x - \frac{x_n + x_m}{2}\| \geq \delta$. Therefore $\|x_n - x_m\|^2 \leq 2\|x_n - x\|^2 + 2\|x_m - x\|^2 - 4\delta^2$. As m and n increase, the right hand side of the last inequality approaches zero which implies that the sequence $\{x_k\}$ is Cauchy and hence has a limit $x_o \in L$. This also shows that the inf is attained by x_o and $\delta = d(x,L) = \min \{\|x - y\| : y \in L\} = \|x - x_o\|$. To show uniqueness suppose that both x_o and x_1 have the above property. Using the parallelogram law we get $\|x_o - x_1\|^2 = 2\|x_o - x\|^2 + 2\|x_1 - x\|^2 - 4\|x - \frac{x_o + x_1}{2}\|^2$. Noting that $\|x_o - x\| = \|x_1 - x\| = \delta$ and $\|x - \frac{x_o + x_1}{2}\| \geq \delta$ we get $\|x_o - x_1\|^2 \leq 0$ which implies $x_o = x_1$. Now we show that x_o is the unique minimizing point if and only if $x - x_o \in L^\perp$. Suppose that $x - x_o \in L^\perp$. Then for $y \in L$ we have $\|x - y\|^2 = \|x - x_o + x_o - y\|^2 = \|x - x_o\|^2 + \|x_o - y\|^2 + 2\langle x - x_o, x_o - y\rangle$. But $x_o - y \in L$ since both x_o and y are in L. Since $x - x_o \in L^\perp$ then $\langle x - x_o, x_o - y\rangle = 0$. So for $x_o \neq y$ we get

$\|x - y\|^2 = \|x - x_o\|^2 + \|x_o - y\|^2 > \|x - x_o\|^2$ which means that x_o is the unique minimizing point. Conversely assume that x_o is the unique minimizing point and by contradiction assume that $x - x_o \notin L^{\perp}$. This means that there is some $y \in L$ such that $\langle x - x_o, y - x_o\rangle > 0$. Now consider $y_\alpha = \alpha y + (1 - \alpha)x_o$ where $\alpha \in (0,1)$. Obviously $y_\alpha \in L$ and we have

$$\|x - y_\alpha\|^2 = \|x - x_o\|^2 + \|x_o - y_\alpha\|^2 - 2\langle x - x_o, y_\alpha - x_o\rangle$$

$$= \|x - x_o\|^2 + \alpha^2\|y - x_o\|^2 - 2\alpha\langle x - x_o, y - x_o\rangle .$$

For α sufficiently small we must have $\alpha^2\|y - x_o\|^2 - 2\alpha\langle x - x_o, y - x_o\rangle < 0$. Therefore $\|x - y_\alpha\|^2 < \|x - x_o\|^2$ for α sufficiently small, contradicting the hypothesis that x_o is a minimizing point. This completes the proof.

It is worthwhile illustrating the above Theorem by a simple example. Consider the following subspace L in E_2:$L = \{(x,y):x + y = 0\}$. We want to find the projection of the point $x = (1,2)$ on L, i.e., we want to find the unique point x_o in L with minimum distance from (1,2) as in Figure 1.2

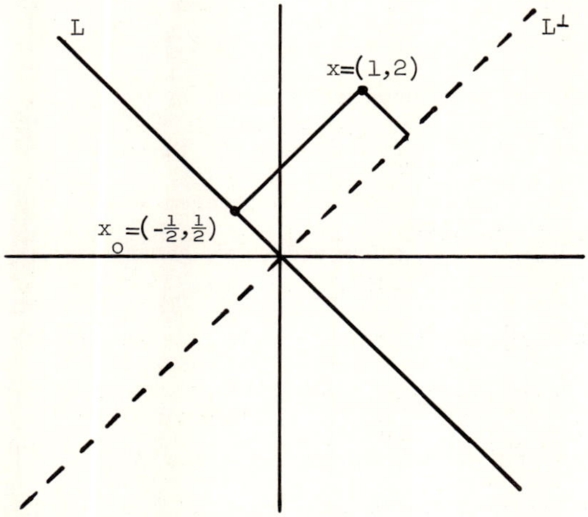

Figure 1.2

An Illustration of the Project Theorem

One way to find $x_o = (\alpha,\beta)$ is to minimize $(1 - \alpha)^2 + (2 - \beta)^2$ where $\alpha + \beta = 0$.

A necessary condition is that $-2(1-\alpha) + 2(2+\alpha) = 0$, i.e., $\alpha = -\frac{1}{2}$ and $\beta = \frac{1}{2}$. Therefore $x_o = (-\frac{1}{2},\frac{1}{2})$ as is also clear from the picture. We can also make use of the characterization $x - x_o \in L^\perp$ to find x_o. Noting that $L^\perp = \{(x,y): x - y = 0\}$ then we get $1 - \alpha = 2 - \beta$ and $\alpha + \beta = 0$ which again gives us the point $(\alpha,\beta) = (-\frac{1}{2},\frac{1}{2})$.

From the above Theorem it follows that given any L in E_n one can project any point x onto L to get x_o with the remainder $x - x_o$ being in $L^\perp$. Since x_o is unique then in essence we developed a decomposition of E_n into L and $L^\perp$, i.e., $E_n = L + L^\perp$. But since L and $L^\perp$ have only the origin in common then E_n can be represented as the direct sum of L and $L^\perp$, i.e., $E_n = L \oplus L^\perp$. x_o is called the orthogonal projection (or simply the projection) of x onto L. Actually the result is symmetric since the projection of x onto $L^\perp$ is indeed $x - x_o$. This follows from the fact that $L = L^{\perp\perp}$. These results are summarized at Theorem 1.3.4 below.

1.3.4 Theorem. Let L be a linear subspace in E_n. Then $E_n = L \oplus L^\perp$ and $L = L^{\perp\perp}$.

Proof: We have seen that given any $x \in E_n$, x can be represented as $x = x_o + y_o$ when $x_o \in L$ and $y_o \in L^\perp$. In order to show that $E_n = L \oplus L^\perp$ we need to show that this representation is unique. So let $x = x_1 + y_1$ with $x_1 \in L$ and $y_1 \in L^\perp$. Hence we get $0 = (x_o - x_1) + (y_o - y_1)$ where $x_o - x_1 \in L$ and $y_o - y_1 \in L^\perp$. Multiplying by $x_o - x_1$ and noting that $\langle x_o - x_1, y_o - y_1\rangle = 0$ we get $\|x_o - x_1\|^2 = 0$, i.e., $x_o = x_1$. Similarly, we conclude that $y_o = y_1$ upon multiplying by $y_o - y_1$. To complete the proof we need to show that $L = L^{\perp\perp}$. It is immediate that $L \subset L^{\perp\perp}$. To show the converse inclusion let $x \in L^{\perp\perp}$. By the projection theorem $x = x_o + y_o$ with $x_o \in L$ and $y_o \in L^\perp$. But since $L \subset L^{\perp\perp}$ then $x_o \in L^{\perp\perp}$ and so we get $x - x_o = y_o \in L^{\perp\perp} \cap L^\perp$. This implies that $x - x_o = 0$ since $L^{\perp\perp} \cap L^\perp = \{0\}$. This shows that $x \in L$ and so $L^{\perp\perp} \subset L$.

The following result is a consequence of Theorem 1.3.4 above. Theorem 1.3.5 states that the dimensions of a linear subspace and its orthogonal complement add up to n.

1.3.5 Theorem. Let L be a nonempty linear subspace in E_n. Then $\dim L + \dim L^\perp = n$.

Proof: Let A and B be bases of L and $L^\perp$ with k and ℓ elements respectively. The reader may note that the elements in A and B form a linearly independent set. Also by the projection theorem any $x \in E_n$ can be represented by $x = x_o + y_o$ with

$x_o \in L$ and $y_o \in L^{\perp}$. But x_o and y_o can be represented as linear combinations of vectors in A and B respectively. Hence $x \in L(A,B)$, i.e., x in a linear combination of the vectors in A and/or B. This shows that A and B combined form a basis of E_n and hence by the corollary to Theorem 1.2.3 we must have $k + \ell = n$.

As a useful application of this Theorem we consider the following example. Consider the set $H = \{x: \langle x,p \rangle = 0\}$ where p is a nonzero vector in E_n. Clearly H is a subspace. Actually H is a hyperplane which passes through the origin and having normal vector p. By definition of $H^{\perp}$ and the construction of H it is obvious that $H^{\perp} = \{\lambda p: \lambda \text{ is real}\}$. From Theorem 1.3.5 it follows that dim H = n - 1 since $H^{\perp}$ is spanned by the single point p. This shows that any hyperplane through the origin has dimension n - 1.

1.4. Affine Manifolds

In many cases we will be interested in translates of subspaces. These are called affine manifolds or simply manifolds. They are referred to by other authors as affine sets, linear varieties, and flats.

1.4.1 Definition. A set M in E_n is called an affine manifold if $M = x + L$ where L is a linear subspace in E_n.

Since the origin belongs to any nonempty linear subspace then $M = x + L$ implies that $x \in M$. It is obvious from the above definition that every subspace is an affine manifold but not conversely. Actually every affine manifold that contains the origin is a linear subspace.

Figure 1.3 shows an example of an affine manifold, namely a line in E_2 not passing by the origin.

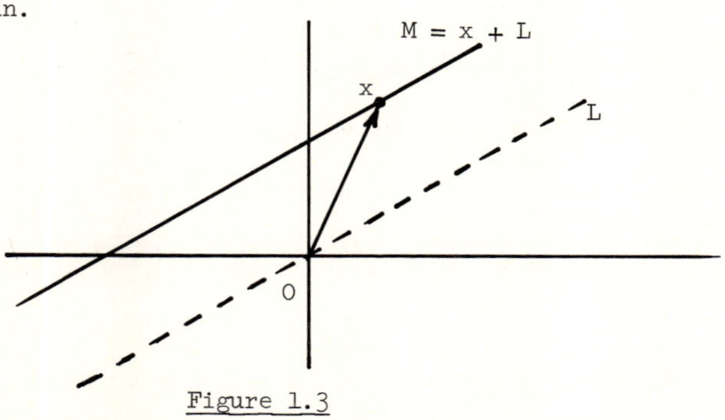

Figure 1.3

An Example of an Affine Manifold

The following lemma shows that an affine manifold cannot be a translate of two distinct subspaces. In other words each affine manifold is parallel to a unique linear subspace.

1.4.2 Lemma. If $M = x_1 + L_1 = x_2 + L_2$ then $L_1 = L_2$.

Proof: We will show that $L_1 \subset L_2$ and by symmetry $L_2 \subset L_1$. $x_1 + x \in x_2 + L_2$ for each $x \in L_1$ and in particular $x_1 - x_2 \in L_2$ by letting $x = 0$. Since $(x_1 - x_2) + x \in L_2$ and $x_1 - x_2 \in L_2$ then $x \in L_2$, i.e., $L_1 \subset L_2$. This completes the proof.

We may note that no claim is made that $x_1 = x_2$. As a matter of fact M can be represented as the sum of its parallel subspace and any point in M. This indicates that $L = M - M = \{x - y : x, y \in M\}$.

As in the case of Lemma 1.1.3, the following can easily be proved:

1.4.3 Lemma. Let M_1 and M_2 be affine manifolds in E_n. Then $M_1 + M_2$ and $M_1 \cap M_2$ are affine manifolds.

The following Theorem gives an important characterization of affine manifolds. We will show that M is a manifold if and only if given any two points in M then the entire line passing through these points must belong to M.

1.4.4 Theorem. Let M be an arbitrary set in E_n. Then M is an affine manifold if and only if $x_1, x_2 \in M$ imply that $\lambda x_1 + (1 - \lambda)x_2 \in M$ for each $\lambda \in E_1$.

Proof: Suppose that M is an affine manifold, i.e., $M = x + L$ where L is the subspace parallel to M and x any point in M. Let x_1 and x_2 be two points in M, i.e., $x_1 = x + y_1$ and $x_2 = x + y_2$ with $y_1, y_2 \in L$. Now

$$\lambda x_1 + (1 - \lambda)x_2 = x + (\lambda y_1 + (1 - \lambda)y_2) \in x + L = M$$

for each $\lambda \in E_1$, i.e., the entire line through x_1 and x_2 lines in M. Conversely suppose that this property holds. We need to show that M is an affine manifold. Let $y_1, y_2 \in L$, i.e., $y_1 = x_1 - x$ and $y_2 = x_2 - x$ with $x_1, x_2 \in M$. We desire to show that $\lambda_1 y_1 + \lambda_2 y_2 \in L$ for each $\lambda_1, \lambda_2 \in E_1$. But

$$\lambda_1 y_1 + \lambda_2 y_2 = (\lambda_1 + \lambda_2)\left[\frac{\lambda_1}{\lambda_1 + \lambda_2} x_1 + \frac{\lambda_2}{\lambda_1 + \lambda_2} x_2 - x\right] = (\lambda_1 + \lambda_2)(y - x)$$

where $y = \frac{\lambda_1}{\lambda_1 + \lambda_2} x_1 + \frac{\lambda_2}{\lambda_1 + \lambda_2} x_2 \in M$ by assumption. Since x and $y \in M$ then we get

$$\lambda_1 y_1 + \lambda_2 y_2 = (\lambda_1 + \lambda_2)(y - x) = (\lambda_1 + \lambda_2)y + (1 - \lambda_1 - \lambda_2)x - x = \hat{y} - x$$

with an obvious definition of $\hat{y} \in M$. This shows that $\lambda_1 y_1 + \lambda_2 y_2 \in M - x = L$ and hence L is indeed a linear subspace. Therefore $M = x + L$ is indeed an affine manifold and the proof is complete.

Needless to say the above Theorem can be equivalently written in terms of more than two points, i.e., M is an affine manifold if and only if $x_1, x_2, \ldots, x_k \in M$ imply that $\sum_{i=1}^{k} \lambda_i x \in M$ where $\lambda_i \in E_1$, $i = 1,2,\ldots,k$ and $\sum_{i=1}^{k} \lambda_i = 1$.

As in the case of a linear space one can generate an affine manifold starting with an arbitrary set S. The reader may review Definition 1.1.4 before proceeding to Definition 1.4.5 below.

1.4.5 Definition. Let S be an arbitrary set in E_n. The affine manifold spanned (generated) by S and denoted by M(S) is given by $M(S) = \{x : x = \sum_{i=1}^{k} \lambda_i x_i,\ x_i \in S,$

$\lambda_i \in E_1,\ \sum_{i=1}^{k} \lambda_i = 1,\ k \geq 1\}$. A set S is said to span (generate) an affine manifold M if $M = M(S)$.

It is clear that for any S, M(S) of the above definition is indeed a manifold. For example let $S = \{(1,0), (1,1)\}$. Then $M(S) = \{(1,y) : y \in E_1\}$. The parallel subspace of M(S) is $L = \{(0,y) : y \in E_1\}$. This is not to be confused with $L(S) = E_2$.

The following lemma which can be easily proved asserts that M(S) is the smallest affine manifold that contains S, a result comparable with Lemma 1.1.5.

1.4.6 Lemma. M(S) is the smallest affine manifold that contains S.

Corollary.

(i) M(S) is the intersection of all affine manifolds containing S.

(ii) S is an affine manifold if and only if $M(S) = S$.

We now give a definition of the dimension of an affine manifold and of an arbitrary set.

1.4.7 Definition. The dimension of an affine manifold M is dim M = dim L where L is the subspace parallel to M. The dimension of an arbitrary set S is dim S = dim M(S).

In other words if we have an arbitrary set S then we first find the affine manifold M(S) generated by it and the parallel subspace L. The dimension of S is then the dimension of L. For example let $S = \{(x,y): 1 \leq x \leq 2,\ x + y = 3\}$ as in Figure 1.4. Then $M(S) = \{(x,y): x + y = 3\}$ and $L = \{(x,y): x + y = 0\}$. It is clear that L is generated by the point (1, -1) and so dim S = dim M(S) = dim L = 1.

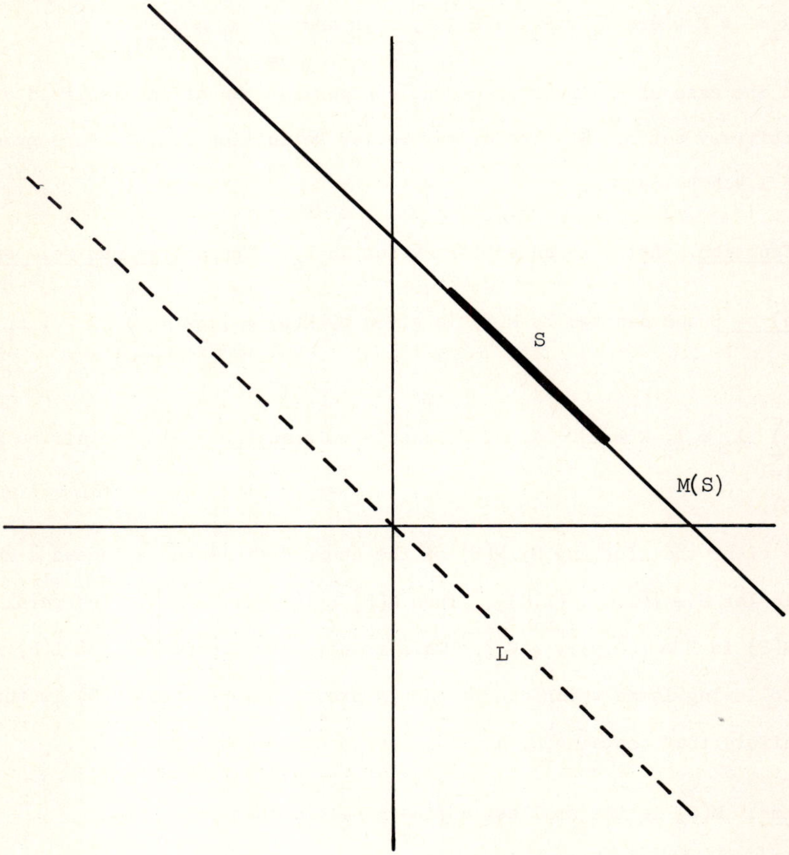

Figure 1.4

An Example of a Dimension of a Set

CHAPTER 2

CONVEX SETS

In the last chapter we discussed the properties of a subspace L. A subspace L had the property that given any two points $x_1, x_2 \in L$, then $\lambda_1 x_1 + \lambda_2 x_2 \in L$ for each $\lambda_1, \lambda_2 \in E_1$. In the above definition if we restrict λ_1 and λ_2 to satisfy $\lambda_1, \lambda_2 > 0$ and $\lambda_1 + \lambda_2 = 1$, we get the definition of a convex set. The above statement can be restated as: the line segment joining any two points in a convex set is in the set. Convex sets is the subject of this chapter. On the other hand, if we restrict λ_1 and λ_2 to be semipositive, then we obtain a cone which is discussed briefly below and in detail in Chapter 3.

2.1. Convex Cones, Convex Sets, and Convex Hulls

2.1.1 Definition. Let C be a subset of E_n. C is said to be a convex cone with vertex zero if x_1 and $x_2 \in C$ imply that every semipositive combination of x_1 and x_2 belong to C. K is called a convex cone with vertex x_o if $K = x_o + C$ where C is a convex cone with vertex zero.

It is to be understood that the vertex of a cone is the origin, of course unless otherwise stated. From the above definition a convex cone with a nonzero vertex is a translation of a unique convex cone with a zero vertex.

Many authors define a convex cone to be a set that contains nonnegative linear combinations of its elements. This will imply that the origin must always belong to the cone (with zero vertex). Here we choose semipositive combinations which gives us the additional flexibility of being able to consider convex cones which do not contain their vertices.

From the above definition it is clear that if x_1, x_2 belong to a convex cone C with zero vertex then so does $\lambda_1 x_1 + \lambda_2 x_2$ for all nonzero nonnegative vectors (λ_1, λ_2). Likewise, $x_o + \lambda_1(x_1 - x_o) + \lambda_2(x_2 - x_o)$ belong to a convex cone K with vertex x_o for each semipositive (λ_1, λ_2) whenever x_1 and x_2 are members of K.

Given an arbitrary set S in E_n, one can construct the minimum convex cone containing S by taking semipositive linear combinations of points in S. This leads to the following definition.

2.1.2 Definition. Let S be an arbitrary set in E_n. C(S) is called the convex cone of S if it consists of all semipositive linear combinations of points in S, i.e.,

$$C(S) = \left\{x : x = \sum_{i=1}^{k} \lambda_i x_i,\ x_i \in S,\ \lambda \text{ is semipositive},\ k \geq 1\right\}.$$

For example, consider the set $S = \{(2,1), (1,2), (2,2)\}$. Then $C(S) = \{(x,y) : (x,y) \neq (0,0),\ 0 \leq \frac{1}{2}x \leq y \leq 2x\}$. Note that the origin is excluded from C(S). Also note that C(S) could have been generated only with the points (1,2) and (2,1), since (2,2) can be represented as a semipositive linear combination of (1,2) and (2,1).

We will now give the definition of a convex set. It may be noted that a convex cone is indeed a special case of a convex set.

2.1.3 Definition. A set X in E_n is said to be convex if $x_1, x_2 \in X$ imply that $\lambda x_1 + (1 - \lambda) x_2 \in X$ for each $\lambda \in (0,1)$.

It may be noted that without additional generality we could have used $k > 2$ points instead of two points in the above definition, i.e., x is a convex set if and only if $x_1, x_2, \ldots, x_k$ imply that $\sum_{i=1}^{k} \lambda_i x_i \in X$ for all $\lambda_i > 0$, $i = 1, 2, \ldots, k$ with $\sum_{i=1}^{k} \lambda_i = 1$.

Again, starting with an arbitrary set S one can find the minimum convex set that contains S. This new set is referred to as the convex hull of S and is denoted by H(S). Here we form the convex combinations of points in S. This gives rise to the following definition.

2.1.4 Definition. Let S be an arbitrary subset of E_n. H(S) is called the convex hull of S if it consists of all convex combinations of S, i.e.,

$$H(S) = \left\{x : x = \sum_{i=1}^{k} \lambda_i x_i,\ x_i \in S,\ \lambda_i > 0,\ i = 1,2,\ldots,k,\ \sum_{i=1}^{k} \lambda_i = 1,\ k \geq 1\right\}$$

From Definitions 2.1.3 and 2.1.4 the following Lemma follows readily.

2.1.5 Lemma. Let S be an arbitrary set in E_n. Then H(S) is the smallest convex set which contains S. Indeed H(S) is the intersection of all convex sets containing S.

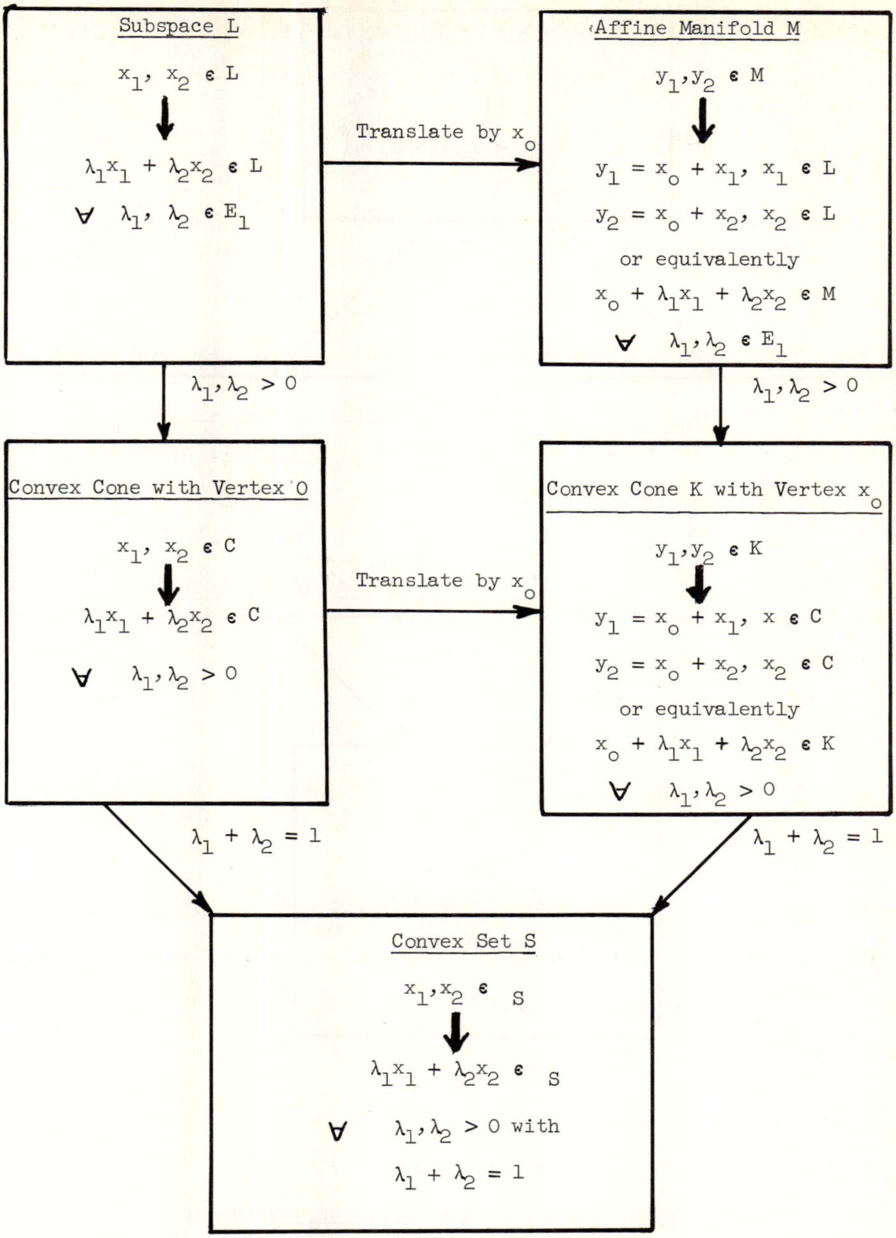

Figure 2.1

Relationships among subspaces, affine manifolds, convex cones, and convex sets

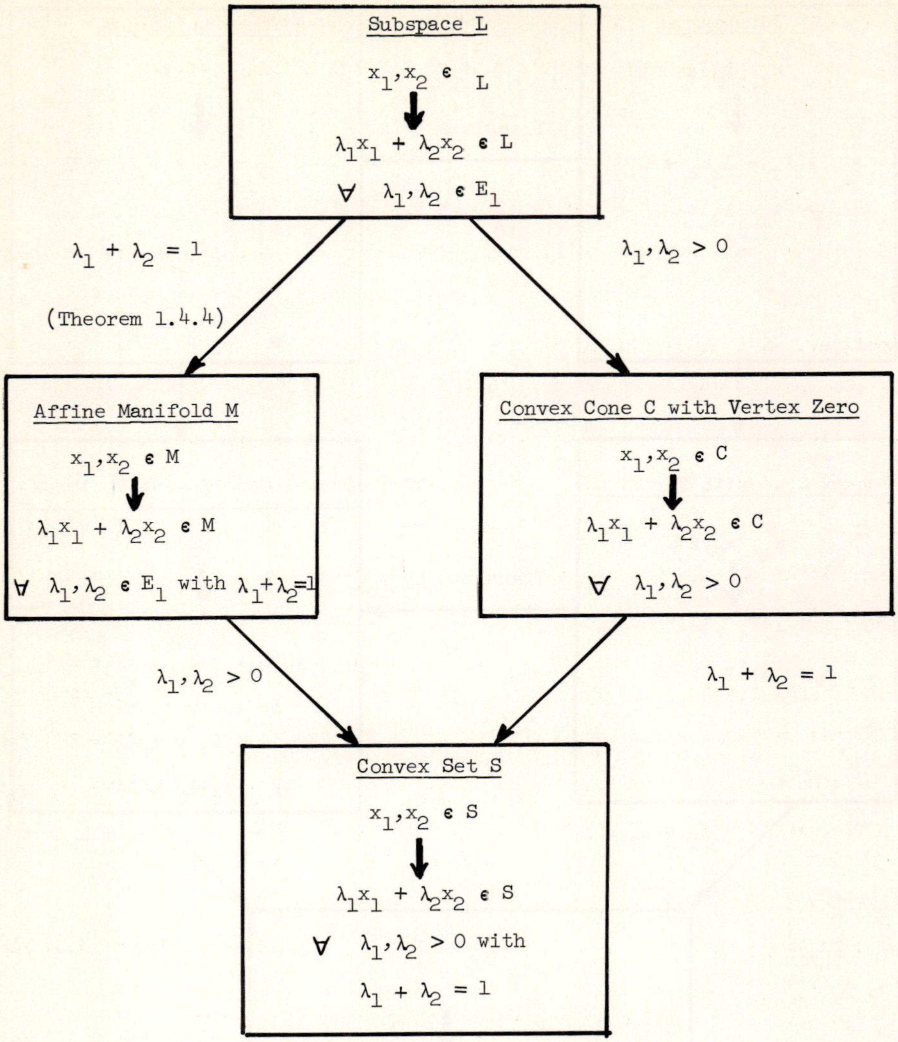

Figure 2.2

Relationships among subspaces, affine manifolds, convex cones, and convex sets

It is clear from the definition of a convex hull that $H(S_1 \cap S_2) \subset H(S_1) \cap H(S_2)$. The other inclusion does not hold necessarily. For example let $S_1 = \{(x,y): y = 1, 0 < x < 1\} \cup \{x = 1, 0 < y < 1\}$ and $S_2 = \{(x,y): x > 0, y > 0, x + y = 1\}$. It is clear that $S_1 \cap S_2$ is empty and so is $H(S_1 \cap S_2)$ but on the other hand $H(S_1) \cap H(S_2) = S_2$ and so $H(S_1 \cap S_2) \neq H(S_1) \cap H(S_2)$.

So far we have developed a heirarchy of sets which have various properties involving their linear combinations, semipositive linear combinations, and convex combinations. The implications between subspaces, affine manifolds, convex cones, and convex sets (convex hulls) are summarized in Figure 2.1. Alternatively, the relationships can be expressed as in Figure 2.2 using Theorem 1.4.4.

As we have mentioned earlier, one can generate the smallest linear subspace, affine manifold, convex cone, and convex set containing an arbitrary set. The relationship among the various convex sets generated by an arbitrary set is depicted in Figure 2.3.

2.2 Carathéodory Type Theorems

In this section we develop several theorems concerning the dimension of convex sets and convex cones. But first some properties of some special convex sets, namely simplices, need be discussed.

2.2.1 Definition. The convex hull of a finite number of points $a_0, a_1, a_2, \ldots, a_k$ is called a polytope. If $a_1 - a_0, a_2 - a_0, \ldots, a_k - a_0$ are linearly independent then $H(a_0, a_1, \ldots, a_k)$ is called a simplex with vertices $a_0, a_1, \ldots, a_k$.

The reader should note that a simplex in E_n cannot have more than $n + 1$ vertices. This follows from Theorem 1.2.3 since E_n does not have $n + 1$ independent vectors.

Let us consider the following points in E_2, $a_0 = (0,1)$, $a_1 = (1,1)$, $a_2 = (1,2)$. It is obvious that $H(a_0, a_1, a_2)$ is indeed a simplex since $a_1 - a_0 = (1,0)$ and $a_2 - a_0 = (1,1)$ are independent. It is also clear that any point in $H(a_0, a_1, a_2)$ other than a_0, a_1, and a_2 can be written as a convex combination of distinct points in $H(a_0, a_1, a_2)$. This distinction between the vertices a_0, a_1, and a_2 and other points in $H(a_0, a_1, a_2)$ motivates the following definition of extreme points of a convex sets. Indeed the vertices of a simplex are extreme points of it.

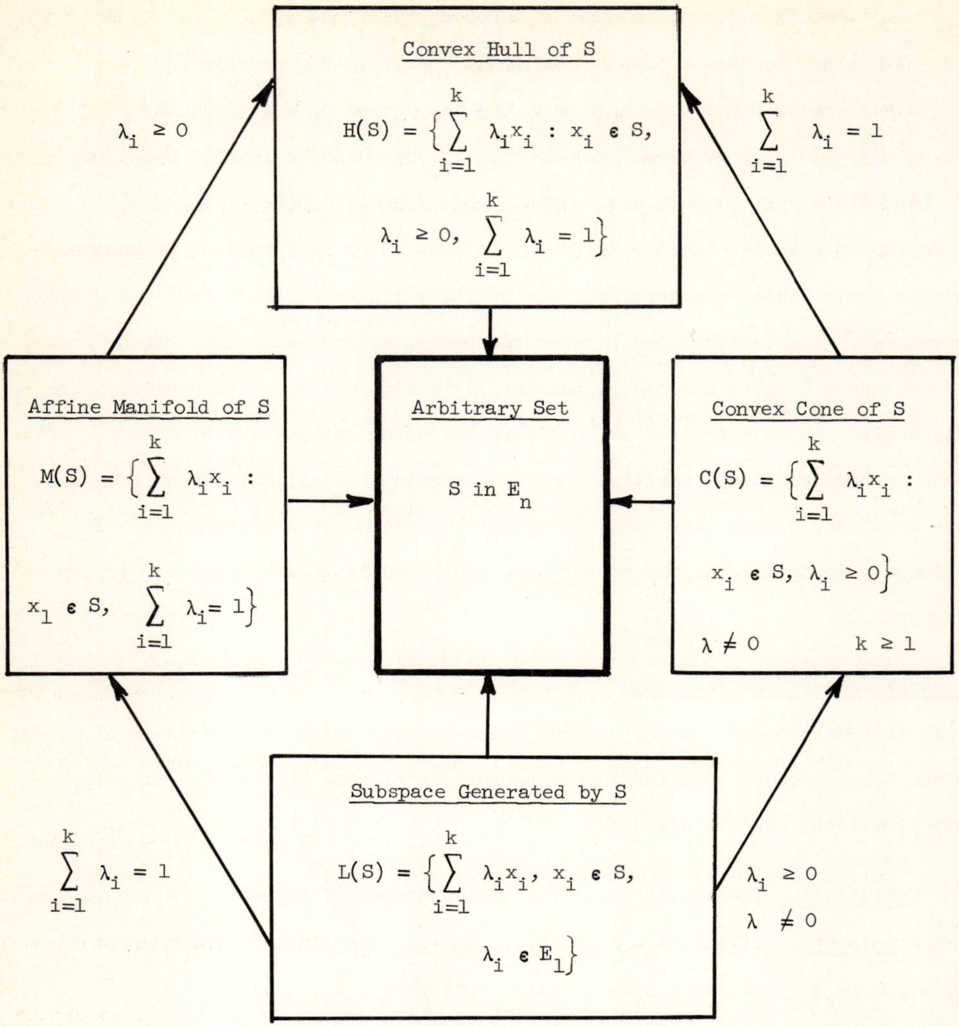

Figure 2.3

Relationship Among Sets Generated by an Arbitrary Set S

(Set at head of the arrow is contained into set at the foot of the arrow)

2.2.2 Definition. A point x of a convex set S is called an extreme point of X if it cannot be written as a convex combination of two distinct points in X.

For example consider the set $S = \{(x,y) : x^2 + y^2 \le 1\}$. All points (x,y) with $x^2 + y^2 = 1$ are extreme points. Going back to the simplex example discussed earlier we observe that the points a_o, a_1, and a_2 are indeed extreme points of $H(a_o, a_1, a_2)$. In general the vertices of a simplex are extreme points. This result is given below.

2.2.3 Lemma. Let $H(a_o, a_1,\ldots,a_k)$ be a simplex with vertices $a_o, a_1,\ldots,a_k$. Then each a_i is a extreme point of $H(a_o, a_1,\ldots,a_k)$.

Proof: We first show that a_o is an extreme point of $H(a_o, a_1,\ldots,a_k)$. By contradiction assume that this is not the case, i.e., there is $x \neq y$ in $H(a_o, a_1,\ldots,a_k)$ such that $a_o = \lambda x + (1 - \lambda)y$ where $\lambda \in (0,1)$. But since both x and y belong to $H(a_o, a_1,\ldots,a_k)$ then $x = \sum_{i=0}^{k} \lambda_i a_i$ and $y = \sum_{i=0}^{k} \mu_i a_i$ where $\lambda_i, \mu_i \ge 0$ and

$\sum_{i=0}^{k} \lambda_i = \sum_{i=0}^{k} \mu_i = 1$. Note that neither λ_o nor μ_o is equal to 1 because this would then imply that $x = y = a_o$ which violates the requirement $x \neq y$. Therefore we get

$a_o = \lambda \sum_{i=0}^{k} \lambda_i a_i + (1 - \lambda) \sum_{i=0}^{k} \lambda_i a_i = \sum_{i=0}^{k} (\lambda\lambda_i + (1 - \lambda)\mu_i)a_i$. This shows that

$\sum_{i=0}^{k} [\lambda\lambda_i + (1 - \lambda)\mu_i]\,(a_i - a_o) = 0$ and hence $\sum_{i=1}^{k} [\lambda\lambda_i + (1 - \lambda)\mu_i]\,(a_i - a_o) = 0$.

Not all $\lambda\lambda_i + (1 - \lambda)\mu_i$ are zero because λ_o and μ_o cannot be equal to 1 as argued above. But this violates the independence of $a_1 - a_o,\ldots,a_k - a_o$. Similarly $a_1, a_2,\ldots,a_k$ can be shown to be extreme points of $H(a_o, a_1,\ldots,a_k)$.

Now let us ask the question can we find a simplex with different collection of vertices? In other words is it possible to find a set S which is the convex combination of $a_o, a_1,\ldots,a_k$ and meanwhile is the convex combination of $b_o, b_1,\ldots,b_m$ with $a_1 - a_o,\ldots,a_k - a_o$ being independent and $b_1 - b_o,\ldots,b_m - b_o$ being independent? The answer is clearly no, unless of course $m = k$ and $\{a_o, a_1,\ldots,a_k\} = \{b_o, b_1,\ldots,b_m\}$. This intuitively obvious fact follows from the above lemma. Consider b_o which is a vertex of the simplex and hence $b_o = \{a_o, a_1,\ldots,a_k\}$. Similarly $b_i \in \{a_o, a_1,\ldots,a_k\}$

for each i and so $\{b_o, b_1, \ldots, b_m\} \subset \{a_o, a_1, \ldots, a_k\}$. By symmetry we get $\{b_o, b_1, \ldots, b_m\} = \{a_o, a_1, \ldots, a_k\}$. This shows that the notion of a vertex of a simplex is well defined as one expects.

The following lemma shows that a simplex having m + 1 vertices has dimension m.

2.2.4 Lemma. Let A be a simplex with vertices $a_o, a_1, \ldots, a_m$. Then dim A = m.

Proof: dim A = dim M(A) = dim L where L is the subspace parallel to M(A). It is clear that $M(A) = \left\{\sum_{i=0}^{m} \lambda_i a_i : \lambda_i \in E_1, \sum_{i=0}^{m} \lambda_i = 1\right\}$. We will now show that $L = \left\{\sum_{i=1}^{m} \lambda_i(a_i - a_o) : \lambda_i \in E_1, i = 1,2,\ldots,m\right\}$. Let x ∈ M(A), i.e., $x = \sum_{i=0}^{m} \lambda_i a_i$ with $\sum_{i=0}^{m} \lambda_i = 1$. So $x = a_o + (\lambda_o - 1)a_o + \sum_{i=1}^{m} \lambda_i a_i = a_o - \sum_{i=1}^{m} \lambda_i a_o + \sum_{i=1}^{m} \lambda_i a_i = a_o + \sum_{i=1}^{m} \lambda_i(a_i - a_o)$. Conversely let $x = a_o + \sum_{i=1}^{m} \lambda_i(a_i - a_o)$ then $x = \sum_{i=0}^{m} \lambda_i a_i$ with $\lambda_o = 1 - \sum_{i=1}^{m} \lambda_i$ and $\sum_{i=0}^{m} \lambda_i = 1$. This shows that L is indeed given by $\left\{\sum_{i=1}^{m} \lambda_i(a_i - a_o) : \lambda_i \in E_1, i = 1,2,\ldots,m\right\}$. But this also shows that dim L = m since $a_1 - a_o, a_2 - a_o, \ldots, a_m - a_o$ are independent and span L. This completes the proof.

As a matter of fact any given point in M(A) is expressible in a unique fashion as a linear combination of $a_o, a_1, \ldots, a_m$ with the coefficients summing to one. To show this let x ∈ M(A) and suppose that $x = \sum_{i=0}^{m} \lambda_i a_i = \sum_{i=0}^{m} \mu_i a_i$ with $\sum_{i=0}^{m} \lambda_i = \sum_{i=0}^{m} u_i = 1$. Therefore, $\mu_o = 1 - \sum_{i=1}^{m} \mu_i = \sum_{i=0}^{m} \lambda_i - \sum_{i=1}^{m} \mu_i = \lambda_o + \sum_{i=1}^{m} (\lambda_i - \mu_i)$. This shows that $\lambda_o a_o + \sum_{i=1}^{m} \lambda_i a_i = \lambda_o a_o + \sum_{i=1}^{m} (\lambda_i - \mu_i)a_o + \sum_{i=1}^{m} \mu_i a_i$. Therefore we get

$\sum_{i=1}^{m} (\lambda_i - \mu_i)(a_i - a_o) = 0$ and by the independence assumption we must have

$\lambda_i = \mu_i$, $i = 1,\ldots,m$. This means that $\lambda_o = \mu_o$, $\lambda_1 = \mu_1,\ldots,\lambda_m = \mu_m$. The representation above is thus unique. $\lambda_1,\lambda_1,\ldots,\lambda_m$ are called the barycentric coordinates of the point x.

Now let us consider the simplex discussed earlier. The vertices a_o, a_1, and a_2 are (0,1), (1,1), and (1,2). It is clear that $M(A) = E_2$ and any point X in E_2 is uniquely expressed by $x = \lambda_o a_o + \lambda_1 a_1 + \lambda_2 a_2$ with $\lambda_o + \lambda_1 + \lambda_2 = 1$. For example the barycentric coordinates of the point (2, -1) are $\lambda_o = -1$, $\lambda_1 = 5$, and $\lambda_2 = -3$. The "centroid" of A, i.e., the point $x_o = \frac{1}{3} a_o + \frac{1}{3} a_1 + \frac{1}{3} a_2 = \left(\frac{2}{3}, \frac{4}{3}\right)$ is called the barycenter of A. Obviously $\lambda_o = \lambda_1 = \lambda_2 = \frac{1}{3}$.

The following lemma relates the dimension of an arbitrary set to the number of independent points in the set.

2.2.5 Lemma. Let S be an arbitrary set in E_n having k + 1 independent vectors. Then dim S $\geq$ k.

Proof: By definition, dim S = dim M(S). Let a_o, $a_1,\ldots,a_k$ be independent vectors in S and let A be the simplex with vertices a_o, $a_1,\ldots,a_k$ (note that $a_1 - a_o$, $a_2 - a_o,\ldots,a_k - a_o$ are independent since a_o, $a_1,\ldots,a_k$ are independent). Clearly $A \subset M(S)$ and therefore dim A $\leq$ dim M(S). This implies that dim A $\leq$ dim S. But by Lemma 2.2.4 dim A = k and hence dim S $\geq$ k.

It should be noted that the result of the above remark cannot be strengthened to dim S being at least k + 1. For example consider the set $S = \{(x,y) : y = 1, 0 \leq x \leq 1\}$. S has the two independent vectors (0,1) and (1,1). However $M(S) = \{(x,y) : y = 1\}$ and dim S = dim M(S) = 1.

The following theorem shows that the dimension of a convex set is equal to the maximum dimension which can possibly be achieved by considering simplices inside the convex set. This result will be used later to show that a nonempty convex set has a nonempty relative interior.

2.2.6 Theorem. Let S be a nonempty convex set in E_n and let $\mathcal{A}$ be the class of all simplices in S. Then dim S = m where $m = \max\{\dim A : A \in \mathcal{A}\}$.

<u>Proof</u>: First note by convexity of S that any simplex with vertices b_o, $b_1, \dots, b_k$ in S is also contained in S. Now let A_m be a maximal simplex in S with vertices a_o, $a_1, \dots, a_m$. Consider the affine manifold $M(A_m)$ generated by A_m. We claim that $S \subset M(A_m)$. By contradiction suppose that this were not the case, i.e., there is an element a_{m+1} in S and not in $M(A_m)$. We now claim that $a_1 - a_o$, $a_2 - a_o, \dots, a_m - a_o$, $a_{m+1} - a_o$ are independent. To show this let $\sum_{i=1}^{m+1} \lambda_i(a_i - a_o) = 0$. If $\lambda_{m+1} = 0$ then by independence of $a_1 - a_o, \dots, a_m - a_o$ it follows that $\lambda_1 = \lambda_2 \dots = \lambda_m = 0$. On the other hand if $\lambda_{m+1} \neq 0$ we get

$$a_{m+1} = a_o + \sum_{i=1}^{m} \frac{\lambda_i}{\lambda_{m+1}} a_o - \sum_{i=1}^{m} \frac{\lambda_i}{\lambda_{m+1}} a_i.$$

By Theorem 1.4.4 it follows that $a_{m+1} \in M(A_m)$ which is impossible. Therefore $\lambda_o = \lambda_2 \dots = \lambda_{m+1} = 0$ and $a_1 - a_o$, $a_2 - a_o, \dots, a_{m+1} - a_o$ are independent. So far we have constructed a simplex in S which has dimension m + 1 violating maximality of A_m. This shows that $S \subset M(A_m)$. Therefore $\dim S \leq \dim M(A_m) = \dim A_m = m$. But on the other hand $S \supset A_m$ and so $\dim S \geq \dim A_m = m$. This shows that $\dim S = m$ and the proof is complete.

We will now turn to Carathéodory type theorems. As we have seen if S is a subset of E_n then the convex hull of S can be obtained by forming all convex combinations of elements of S. Carathéodory type theorems in effect enable one to consider convex combinations involving at most n + 1 points.

The development in the rest of this section is intuitively appealing from a geometrical point of view. An important relationship can be established between elements of a convex cone in E_{n+1} and elements of a convex set in E_n, and also between a subspace in E_{n+1} and an affine manifold in E_n. This is stated precisely in Lemma 2.2.7 below wherein it may be observed that increasing the dimension readily permits the restriction $\Sigma\lambda_i = 1$ to be relaxed, of course, keeping in mind that the restriction still holds but in an implicit form. The lemma will permit some theorems on cones to be applied to convex sets and theorems on subspaces to be applied to affine manifolds.

For notational convenience we will associate with each point $x \in E_n$ a point $\hat{x}$ in E_{n+1} where $\hat{x} = (x,1)$. Also we will denote $\hat{S}$ as the set of $\hat{x}$ when x belongs to S, i.e., $\hat{S} = \{\hat{x} : \hat{x} = (x, 1), x \in S\}$. With this notation, the following lemma is obvious.

2.2.7 Lemma

(i) $x \in H(S)$ if and only if $(x, 1) \in C(\hat{S})$, i.e., $x \in H(S)$ if and only if $\hat{x} \in C(\hat{S})$.

(ii) $x \in M(S)$ if and only if $(x, 1) \in L(\hat{S})$, i.e., $x \in M(S)$ if and only if $\hat{x} = L(\hat{S})$.

Figure 2.4 may be helpful in showing the relationship between H(S) and $C(\hat{S})$. A similar demonstration can be established between M(S) and $L(\hat{S})$.

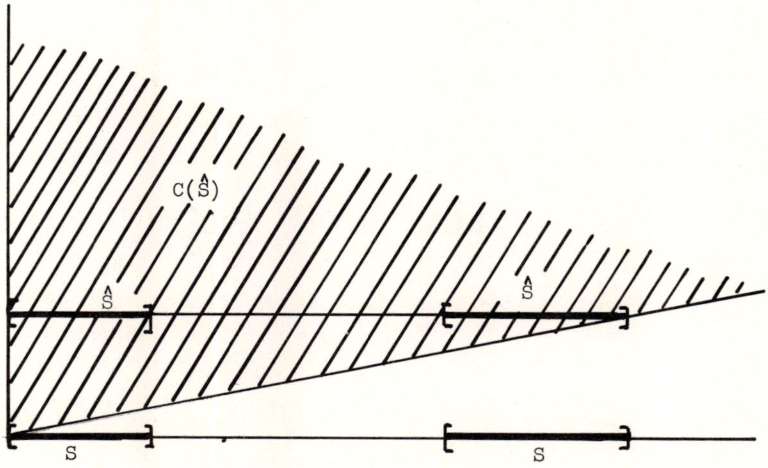

Figure 2.4

Relationship between H(S) and $C(\hat{S})$

The following theorem shows that a point in the subspace generated by a set S can be represented by a linear combination of at most m + 1 points where m = dim S. We need the following lemma first.

2.2.8 Lemma. Let S be a nonempty set in E_n and L be the linear subspace parallel to M(S). Let $x_1, x_2, \ldots, x_m$ be a basis of L. Then $\hat{y}_0, \hat{y}_1, \ldots, \hat{y}_m$ is a basis of $L(\hat{S})$ where x_0 is an arbitrary point in M(S) with $y_0 = x_0$ and $y_i = x_0 + x_i$ for

$i = 1, 2, \ldots, m$ and $\hat{y}_i = (y_i, 1)$ for $i = 0, 1, \ldots, m$.

Proof: Let $(u, v) \in L(\hat{S})$. We first desire to show that (u, v) can be written as a linear combination of $\hat{y}_o, \hat{y}_1, \ldots, \hat{y}_m$. If $v = 0$ then $(u, 0) \in L(\hat{S})$ which means that $(u, 0) = \sum_{i=1}^{k} \mu_i(z_i, 1)$ with $z_i \in S$. Therefore $\sum_{i=1}^{k} \mu_i = 0$ and

$(u, 0) = \sum_{i=1}^{k} \mu_i(z_i - x_o, 1) + \sum_{i=1}^{k} \mu_i(x_o, 0) = \sum_{i=1}^{k} \mu_i(z_i - x_o, 1)$. This shows

that $u \in L$ and hence $u = \sum_{i=1}^{m} \lambda_i x_i$. Therefore $(u, 0) = \sum_{i=1}^{m} \lambda_i(x_o + x_i, 1)$

$+ \lambda_o(x_o, 1) = \sum_{i=0}^{m} \lambda_i \hat{y}_i$ where $\lambda_o = - \sum_{i=1}^{m} \lambda_i$. This shows that $(u, 0) \in L(\hat{y}_o, \hat{y}_1, \ldots,$

$\hat{y}_m)$. We will now consider the case when $v \neq 0$. Since $L(\hat{S})$ is a subspace then $\left(\frac{u}{v}, 1\right) \in L(\hat{S})$ and by Lemma 2.2.7 $\frac{u}{v} \in M(S)$. Therefore $\frac{u}{v} = x_o + \sum_{i=1}^{m} \lambda_i x_i$. Letting

$\lambda_o = 1 - \sum_{i=1}^{m} \lambda_i$ then $\frac{u}{v} = \lambda_o x_o + \sum_{i=1}^{m} \lambda_i(x_o + x_i) = \sum_{i=0}^{m} \lambda_i y_i$. Since $\sum_{i=0}^{m} \lambda_i = 1$

then $\frac{u}{v} \in M(y_o, y_1, \ldots, y_m)$ and by Lemma 2.2.7 $\left(\frac{u}{v}, 1\right) \in L(\hat{y}_o, \hat{y}_1, \ldots, \hat{y}_m)$ and $(u, v) \in L(\hat{y}_o, \hat{y}_1, \ldots, \hat{y}_m)$. To complete the proof we need to show that $\hat{y}_o, \hat{y}_1, \ldots, \hat{y}_m$

are independent. So if we let $\sum_{i=0}^{m} \lambda_i \hat{y}_i = \lambda_o(x_o, 1) + \sum_{i=1}^{m} \lambda_i(x_o + x_i, 1) = (0, 0)$

then we need to show that $\lambda_o = \lambda_1 = \ldots \lambda_m = 0$. From the above equation we get

$\lambda_o = - \sum_{i=1}^{m} \lambda_i$. This shows that $\sum_{i=1}^{m} \lambda_i x_i = 0$ and since $x_1, x_2, \ldots, x_m$ form a basis of L then they are independent and $\lambda_1 = \lambda_2 = \ldots = \lambda_m = 0$. This also shows that $\lambda_o = 0$ and the proof is complete.

The following four theorems show that points in the linear subspace, affine manifold, convex cone, or convex hulls generated by arbitrary sets can be represented as points in the linear subspace, affine manifold, convex cone, or convex hulls generated by a finite number of points from the original set. The number of points is related to the dimension of the set. Actually these theorems can be viewed as Carathéodory type theorems.

2.2.9 Theorem. Let S be a nonempty set in E_n with dimension m. If $x \in L(S)$ then there exists $x_1, x_2, \ldots, x_{m+1} \in S$ such that $x \in L(x_1, x_2, \ldots, x_{m+1})$, i.e.,

$$x = \sum_{i=1}^{m+1} \lambda_i x_i \text{ where } x_i \in S, \lambda_i \in E_1, i = 1, 2, \ldots, m+1.$$

Proof: Since $x \in L(S)$ then $x = \sum_{i=1}^{k} \lambda_i x_i$ where $x_i \in S$, $\lambda_i \in E_1$, $i = 1,2,\ldots,k$. If $k \le m+1$ then the result is immediate. If $k > m+1$ then $x_1, x_2, \ldots, x_k$ are dependent because otherwise by Lemma 2.2.5 dim $S \ge m+1$ contradicting the hypothesis dim $S = m$. Since $x_1, x_2, \ldots, x_k$ are dependent, then without loss of generality let

$$x_k = \sum_{i=1}^{k-1} \mu_i x_i. \text{ Then we get } x = \sum_{i=1}^{k-1} \lambda_i x_i + \lambda_k x_k = \sum_{i=1}^{k-1} (\lambda_i + \lambda_k \mu_i) x_i, \text{ i.e., } x \text{ is}$$

expressed as a linear combination of k-1 points of S. The process is repeated until we reach $m + 1$.

For example let $S = \{(x, y) : x + y) = 1\}$. It is clear that dim $S = 1$ and that any point in $L(S) = E_2$ can be represented as a linear combination of two points in S, e.g., (3, -7) is a linear combination of (2, -1) and (0, 1) with $\lambda_1 = \frac{3}{2}$ and $\lambda_2 = -\frac{11}{2}$. Clearly the representation is not unique, for example, $(3, -7) = \lambda_1(1, 0) + \lambda_2(-2, 3)$ with $\lambda_1 = -\frac{5}{3}$ and $\lambda_2 = -\frac{7}{3}$.

The following theorem, similar to the above theorem, shows that a point in the affine manifold generated by an arbitrary set with dimension m is contained in the affine manifold generated by $m + 1$ points from the original set.

2.2.10 Theorem. Let S be a nonempty set in E_n with dimension m. If $x \in M(S)$ then there exists $x_1, x_2, \ldots, x_{m+1}$ such that $x \in M(x_1, x_2, \ldots, x_{m+1})$.

Proof: Let L be the subspace parallel to M(S), i.e., let $M(S) = b_o + L$ where $b_o \in M(S)$. Let $b_1, b_2, \ldots, b_m$ be a basis of L. By Lemma 2.2.8, $\hat{y}_o, \hat{y}_1, \ldots, \hat{y}_m$ form a basis of $L(\hat{S})$ where $y_o = b_o$ and $y_i = b_o + b_i$ for $i = 1, 2, \ldots, m$, and $\hat{y}_i = (y_i, 1)$ for

$i = 0, 1, \ldots, m$. Letting $x \in M(S)$ it follows by Lemma 2.2.7 that $\hat{x} \in L(\hat{S})$. Since $\hat{y}_o, \hat{y}_1, \ldots, \hat{y}_m$ form a basis of $L(\hat{S})$ then $\hat{x} \in L(\hat{y}_o, \hat{y}_1, \ldots, \hat{y}_m)$. Again by Lemma 2.2.7 we get $x \in M(y_o, y_1, \ldots, y_m) = M(b_o, b_o + b_1, \ldots, b_o + b_m)$.

2.2.11 Theorem. Let S be a nonempty set in E_n and let $C(S)$ be the convex cone generated by S. Suppose that $\dim C(S) = m$. If $x \in C(S)$ then there exist $x_1, x_2, \ldots, x_m$ in S such that $x \in C(x_1, x_2, \ldots, x_m)$.

Proof: First note that $\dim L(S) = \dim C(S)$. Since $x \in C(S)$ then

$x = \sum_{i=1}^{k} \lambda_i x_i$, $x_i \in S$, $\lambda_i > 0$, $i = 1, 2, \ldots, k$. If $k \le m$ then the results obviously holds. If $k > m$ we will show that x can be represented as a positive linear combination of $k - 1$ points. Note that if $k > m$ then $x_1, x_2, \ldots, x_k$ are linearly dependent because otherwise $\dim C(S) = \dim L(S) \ge k > m$ violating the hypothesis of the theorem. Hence there exist scalars $\mu_1, \mu_2, \ldots, \mu_k$, not all zero, such that $\sum_{i=1}^{k} \mu_i x_i = 0$. Without loss of generality we can assume that at least one $\mu_i > 0$.

Let $\frac{\lambda_s}{\mu_s} = \min_i \left\{\frac{\lambda_i}{\mu_i} : \mu_i > 0\right\}$ so that $\bar{\lambda}_i = \left(\lambda_i - \frac{\lambda_s}{\mu_s} \mu_i\right) \ge 0$ for $i = 1, 2, \ldots, k$ with

$\bar{\lambda}_s = 0$. Further $x = \sum_{i=1}^{k} \lambda_i x_i = \sum_{i=1}^{k} \bar{\lambda}_i x_i + \frac{\lambda_s}{\mu_s} \sum_{i=1}^{k} \mu_i x_i = \sum_{\substack{i=1 \\ i \ne s}}^{k} \bar{\lambda}_i x_i$ since $\bar{\lambda}_s = 0$

and $\sum_{i=1}^{k} \mu_i x_i = 0$. Therefore x can be represented as a nonnegative linear combination of $k - 1$ points. We repeat the process until m is reached.

The following theorem shows that any point in the convex hull of a set S can be written as a convex combination of $m + 1$ points of S where $\dim S = m$. In particular then points in $H(S)$ can be written as the convex combination of $n + 1$ points from S.

2.2.12 Theorem. Let S be a nonempty set in E_n with $\dim S = m$. If $x \in H(S)$ then there exist $x_1, x_2, \ldots, x_{m+1}$ in S with $x \in H(x_1, x_2, \ldots, x_{m+1})$.

Proof: Let b_o be an arbitrary point in $M(S)$ and let L be the subspace parallel to $M(S)$. Let $b_1, b_2, \ldots, b_m$ be a basis of L. Then by Lemma 2.2.8, $\hat{y}_o, \hat{y}_1, \ldots, \hat{y}_m$ form a basis of $L(\hat{S})$ where $\hat{y}_i = (y_i, 1)$ for $i = 0, 1, \ldots, m$ and $y_o = b_o$ and $y_i = b_o + b_i$

for i = 1, 2,...,m. Therefore dim $C(\hat{S})$ = dim $L(\hat{S})$ = m + 1. Now let $x \in H(S)$. Then by Lemma 2.2.7 $\hat{x} \in C(\hat{S})$ and by theorem there exist $\hat{x}_1, \hat{x}_2, \ldots, \hat{x}_{m+1}$ in $\hat{S}$ such that $\hat{x} \in C(\hat{x}_1, \hat{x}_2, \ldots, \hat{x}_{m+1})$. Again by Lemma 2.2.7 we must have $x \in H(x_1, x_2, \ldots, x_m)$ where $\hat{x}_i = (\hat{x}_i, 1)$ for i = 1,2,...,m + 1. This completes the proof.

2.3. Relative Interior and Related Properties of Convex Sets

In this section we will talk about interior, relative interior, boundary, and relative boundary of a set. After that we will develop some important theorems which essentially deal with convex sets.

Consider the following definition of an interior point.

2.3.1 Definition. Let S be a nonempty set in E_n. $x_o \in S$ is said to be an interior point of S if there exists an open ball $N_\epsilon(x_o)$ about x_o with radius $\epsilon > 0$ which is contained in S, i.e., $N_\epsilon(x_o) \subset S$. The set of all interior points of S is called the interior of S, and is denoted by int S. S is said to be open if S = int S.

The existence of an interior point is often too strong an assumption in many theorems dealing with optimization problems. Now consider the set $S = \{(x, y) : 0 < x < 2, y = 0\}$. From Definition 2.3.1 above int S = $\emptyset$. However, it seems reasonable to consider all points of S as "interior" points since these points would have satisfied the above definition had we considered S as a set in E_1 rather than E_2. It is precisely to incorporate this point of view that the notion of a relative interior is used. This notion is defined below.

2.3.2 Definition. Let S_1 and S_2 be two sets in E_n with $S_1 \subset S_2$. $x \in S_1$ is said to be an interior point of S_1 relative to S_2 if there exists an $\epsilon > 0$ such that $N_\epsilon(x) \cap S_2 \subset S_1$. S_1 is said to be open relative to S_2 if every point of S_1 is an interior point of S_1 relative to S_2. Now let S be a nonempty set in E_n with $x \in S$. x is said to be in the relative interior of S if x is an interior point of S relative to M(S), the affine manifold of S. S is said to be relatively open if every point of S is in the relative interior of S.

If x is in the relative interior of S then we write $x \in$ riS. From the above definition it is evident that riS = $\{x :$ there is an $\epsilon > 0$ with $N_\epsilon(x) \cap M(S) \subset S\}$. Comparing this with Definition 2.3.1 of the interior of a set it follows that the notion of the interior and the relative interior are equivalent if $M(S) = E_n$; i.e., when S has dimension n.

Now consider the example discussed above, namely let $S = \{(x, y) : 0 < x < 2, y = 0\}$. It is then clear that $M(S) = \{(x, y) : x \in E_1, y = 0\}$ and hence $riS = S$. In essence the notion of relative interior made us consider neighborhoods in the form of intervals rather than disks in E_2. This is shown in Figure 2.5.

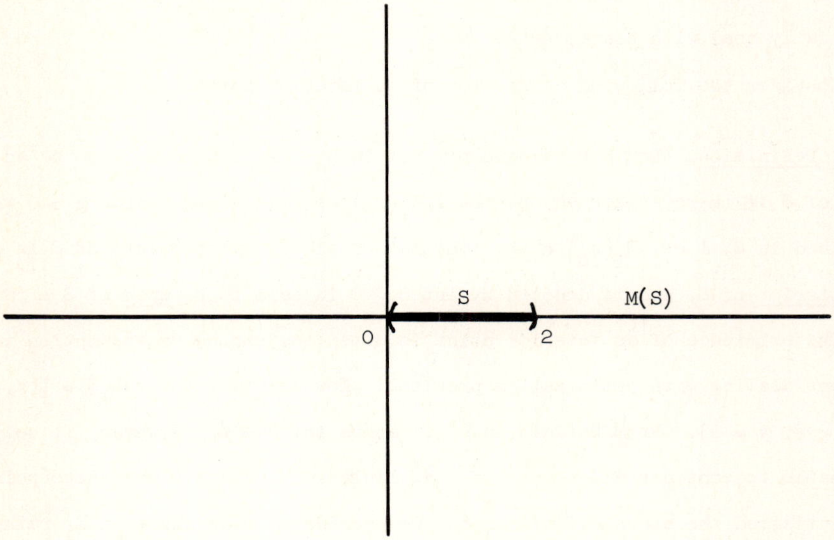

Figure 2.5

Relationship Between Interior and Relative Interior

It should be noted that $S_1 \subset S_2$ implies int $S_1 \subset$ int S_2. However, this result does not hold in general for relative interiors. For example, let $S_1 = \{x : \langle x, \xi \rangle = \alpha\}$ and $S_2 = \{x : \langle x, \xi \rangle \leq \alpha\}$. Here we have $S_1 \subset S_2$ but on the other hand $riS_1 = S_1$ and $riS_2 = \{x : \langle x, \xi \rangle < \alpha\}$, i.e., $riS_1 \not\subset riS_2$. The following is a sufficient condition for $riS_1 \subset riS_2$ to hold as long as $S_1 \subset S_2$.

2.3.3 Lemma. Let $S_1 \subset S_2$ be two sets in E_n. If $M(S_1) = M(S_2)$ then $riS_1 \subset riS_2$.

Proof: Let $x \in riS_1$, i.e., there exists an $\epsilon > 0$ such that $N_\epsilon(x) \cap M(S_1) \subset S_1$. But since $M(S_1) = M(S_2)$ then $N_\epsilon(x) \cap M(S_2) \subset S_1 \subset S_2$ which shows that $x \in riS_2$.

Needless to say that the condition $M(S_1) = M(S_2)$ could have been replaced by $M(S_1) \supset M(S_2)$ since the inclusion $M(S_1) \subset M(S_2)$ always hold as long as $S_1 \subset S_2$.

We will now discuss a special set and show that it indeed has nonempty relative interior. Hopefully this will give the reader some insight into the notion of a relative interior. Let us consider the simplex A in E_n with vertices $a_0, a_1, \ldots, a_m$. We will show that riA is not empty by showing that there is a point $x \in$ riA. Construct the following affine manifolds (see Figure 2.6):

$$M_o = M(a_1, a_2, \ldots, a_m)$$
$$M_m = M(a_o, a_1, \ldots, a_{m-1})$$
$$M_i = M(a_o, a_1, \ldots, a_{i-1}, \ldots, a_m) \qquad i = 1, 2, \ldots, m-1$$

i.e., M_i is the affine manifold generated by all the vertices of A except a_i. Now consider any point $x = \sum_{i=0}^{m} \lambda_i a_i$ where $\lambda_i > 0$ for each i and $\sum_{i=0}^{m} \lambda_i = 1$. Clearly the distance from x to any M_i is positive. This can be shown as follows. Suppose that

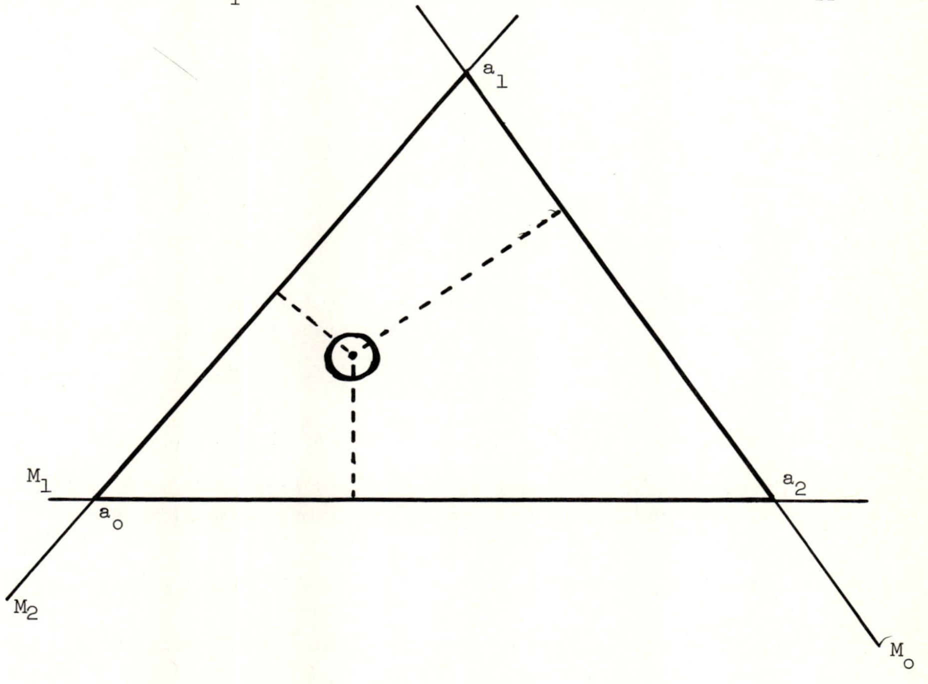

Figure 2.6

Relative Interior of a Simplex

$d(x, M_i) = 0$ for some $i \in \{0, 1, \ldots, m\}$ then x can be represented as $\sum_{j \neq i} \mu_j a_j$ where $\sum_{j \neq i} \mu_i = 1$. Therefore, $\sum_{j=0}^{m} \lambda_j a_j = \sum_{j \neq i} \mu_j a_j$ and hence $\lambda_i a_i = \sum_{j \neq i} (\mu_j - \lambda_j) a_j$. Note that $\mu_j - \lambda_j$ is not zero for each $\mu_j - \lambda_j$ since $\lambda_i \neq 0$. Letting $\mu_j - \lambda_j$ be α_j and noting that $\sum_{j \neq i} \alpha_j = \lambda_i$ then we get $\sum_{j \neq i} \alpha_j (a_j - a_i) = 0$ which violates the independence of $a_j - a_i$ for $j \neq i$. This then shows that $d(x, M_i) > 0$ for each i. Now let D be the minimum of the distances from x to M_i's, i.e., $D = \min\{d(x, M_i) : i = 0, 1, \ldots m\} > 0$. We will show that $x \in \text{ri}A$ by showing that $N_D(x) \cap M(A) \subset A$. In other words we will show that given a point $y \in M(A)$ with $\|x - y\| < D$ then y must also belong to the simplex A. Suppose to the contrary of this desired result there exists some $y \in M(A)$ such that $\|x - y\| < D$ and $y \notin A$. $y \in M(A)$ implies that $y = \sum_{i=0}^{m} \mu_i a_i$ where $\sum_{i=0}^{m} \mu_i = 1$. Since $y \notin A$ then for some j, $\mu_j < 0$. Consider convex combinations of x and y, i.e., consider $x_\alpha = \alpha x + (1 - \alpha) y$ for $\alpha \in (0, 1)$. $x_\alpha = \alpha \sum_{i=0}^{m} \lambda_i a_i + (1 - \alpha) \sum_{i=0}^{m} \mu_i a_i = \sum_{i=0}^{m} (\alpha\lambda_i + (1 - \alpha)\mu_i) a_i$. Since $\lambda_j > 0$ and $\mu_j < 0$ then for some $\alpha_o \in (0, 1)$ we should have $\alpha_o \lambda_j + (1 - \alpha_o)\mu_j = 0$. Therefore, $x_{\alpha_o} = \sum_{i=0}^{m} (\alpha_o \lambda_i + (1 - \alpha_o)\mu_i) a_i$. Since $\sum_{i=0}^{m} \lambda_i = \sum_{i=0}^{m} \mu_i = 1$ it is obvious that $\sum_{\substack{i=0 \\ i \neq j}}^{m} (\alpha_o \lambda_i + (1 - \alpha_o)\mu_i) = 1$. Also x_{α_o} is a linear combination of $a_o, \ldots, a_{j-1}, a_{j+1}, \ldots, a_m$ and hence by definition of M_j, $x_{\alpha_o} \in M_j$. But now consider $\|x - x_{\alpha_o}\| = \|x - \alpha_o x - (1 - \alpha_o) y\| = (1 - \alpha_o)\|x - y\| < (1 - \alpha_o) D < D$. This contradicts the fact that $d(x, M_j) \geq D$ since $x_{\alpha_o} \in M_j$). Therefore, $y \in A$ and hence $x \in \text{ri}A$.

In essence we have shown that the relative interior of the simplex A consists of points of the form $\sum_{i=0}^{m} \lambda_i a_i$ where $\lambda_i > 0$ and $\sum_{i=0}^{m} \lambda_i = 1$. The importance of each λ_i to be strictly positive is not to be overemphasized. Actually it can be shown that if any of the λ_i's is zero then the point is not in the relative interior. It is also interesting to note that picking the λ_i's in a certain fashion will lead to a point which has the same distance from each of the manifolds. This can be done by calculating $d_i = d(a_i, M_i)$. Letting $D = 1/\Sigma(1/d_i)$ and $\lambda_i = \frac{D}{d_i}$. The details of showing that this is indeed the case is left to the reader.

It is worthwhile computing a neighborhood that will work for a given point. In particular we will pick the point equidistant from the manifolds. Consider the manifold in E_2 with vertices (0, 0), (1, 0) and (0, 1) shown in Figure 2.7.

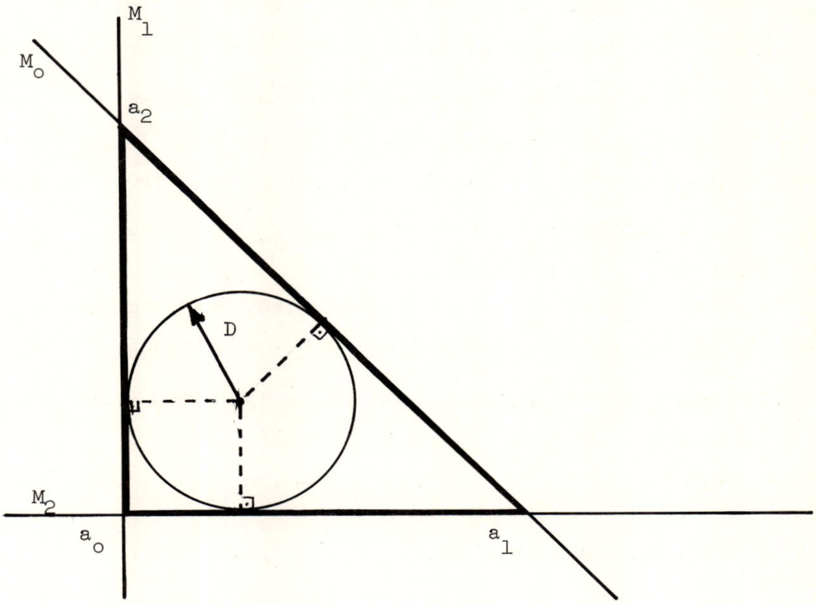

Figure 2.7.

An Example of Relative Interior of a Simplex

Clearly M_1 and M_2 are the y and x axes respectively and $M_0 = \{(x, y) : x + y = 1\}$. Now $d_0 = d(a_0, M_0) = \frac{1}{\sqrt{2}}$ and $d_1 = d_2 = 1$. Therefore, $D = \frac{1}{1+1+\sqrt{2}} = \frac{1}{\sqrt{2}(1+\sqrt{2})}$ and so $\lambda_0 = \frac{1}{1+\sqrt{2}}$, $\lambda_1 = \frac{1}{\sqrt{2}(1+\sqrt{2})}$, and $\lambda_2 = \frac{1}{\sqrt{2}(1+\sqrt{2})}$. Clearly $\lambda_i > 0$ for i = 0, 1, 2 and $\sum_{i=0}^{2} \lambda_i = 1$. Now the point $x = \lambda_0 a_0 + \lambda_1 a_1 + \lambda_2 a_2 = \frac{1}{\sqrt{2}(1+\sqrt{2})}(1, 1)$ is in the relative interior of A with equidistance from M_0, M_1, and M_2, namely $D = \frac{1}{\sqrt{2}(1+\sqrt{2})}$.

From the above discussion and example it is evident that a simplex has a non-empty relative interior. But we know by Theorem 2.2.6 that a convex set contains simplices having the same dimension as the set. This in effect shows that a nonempty convex set indeed has a nonempty relative interior. This result is made more precise by the following theorem. Of course, a parallel result does not hold in general if the relative interior is replaced by the interior.

2.3.4 Theorem. Let S be a nonempty convex set in E_n. Then $\text{ri}S \neq \emptyset$.

Proof: If S consists of one point then $S = \text{ri}S$ and the result is obvious. Now let $\dim S = m \geq 1$. By Theorem 2.2.6 there exists a simplex A with dimension m and vertices $a_o, a_1, \ldots, a_m$ such that $M(A) = M(S)$. By Lemma 2.3.3 we get $\text{ri}A \subset \text{ri}S$. But riA is not empty since it consists of points of the form $\sum_{i=0}^{m} \lambda_i a_i$ with $\lambda_i > 0$ for each i and $\sum_{i=0}^{m} \lambda_i = 1$ and so $\text{ri}S \neq \emptyset$. This completes the proof.

The following theorem shows that the line segment (excluding the end points) joining a point in the relative interior and a point in the closure of a convex set is contained in the relative interior of the convex set. This result along with Theorem 2.3.4 above are of special importance in deriving many other results.

2.3.5 Theorem. Let S be a nonempty convex set in E_n. Let $x_1 \in C\ell S$ and $x_2 \in \text{ri}S$. Then $y = \lambda x_1 + (1 - \lambda)x_2 \in \text{ri}S$ for each $\lambda \in (0, 1)$.

Proof: Since $x_2 \in \text{ri}S$ then there is an $\varepsilon > 0$ such that $N_\varepsilon(x_2) \cap M(S) \subset S$. To prove that $y \in \text{ri}S$ it suffices to show that there exists a neighborhood about y such that its intersection with the affine manifold generated by S is contained in S. In particular we will show that $N_{(1-\lambda)\varepsilon}(y)$ is such a neighborhood. Letting $z \in M(S)$ such that $\|z - y\| < (1 - \lambda)\varepsilon$ we need to show that $z \in S$. We will do this by showing that z can be expressed as a positive linear combination of z_1 and z_2 in S. See Figure 2.8 below. Since $x_1 \in C\ell S$ then $N_\delta(x_1) \cap S \neq \emptyset$ for each $\delta > 0$. Now let $\delta = \frac{(1-\lambda)\varepsilon - \|z-y\|}{\lambda} > 0$ and choose $z_1 \in N_\delta(x_1) \cap S$, i.e., $z_1 \in S$ and $0 < \|z_1 - x_1\| < \frac{(1-\lambda)\varepsilon - \|z-y\|}{\lambda}$. Let $z_2 = \frac{z}{1-\lambda} - \frac{\lambda}{1-\lambda} z_1$. Since $z, z_1 \in M(S)$ and $\frac{1}{1-\lambda} + \frac{-\lambda}{1-\lambda} = 1$ it is obvious that $z_2 \in M(S)$. Furthermore, by definition of y and choice of z_1 we get

$$\|z_2 - x_2\| = \left\| \frac{z-\lambda z_1}{1-\lambda} - x_2 \right\| = \left\| \frac{z-\lambda z_1 - (y-\lambda x_1)}{1-\lambda} \right\| \leq \frac{1}{1-\lambda} (\| z - y\| + \lambda\|x_1 - z_1\|)$$

$< \frac{1}{1-\lambda} (1-\lambda)\varepsilon = \varepsilon$. This shows that $z_2 \in N_\varepsilon(x_2) \cap M(S)$ and hence $z_2 \in X$. Therefore

$z = \lambda z_1 + (1 - \lambda)z_2$ where z_1, $z_2 \in S$ and $\lambda \in (0, 1)$. By convexity of X it then follows that $z \in S$ and the proof is complete.

<u>Corollary 1</u>. Let S be a nonempty convex set in E_n. Let $x_1 \in C\ell\, S$ and $x_2 \in$ int S. Then $y = \lambda x_1 + (1 - \lambda)x_2 \in$ int S for each $\lambda \in (0, 1)$.

<u>Proof</u>: Either int S = riS if dim S = n or else int S = $\emptyset$. In the first case result follows trivially from the theorem and in the second case the corollary does not say anything since we cannot find such an x_2.

<u>Corollary 2</u>. If S is convex then riS and int S are both convex.

<u>Proof</u>: Let $x_1 \in$ riS in the above theorem. int S is convex since int S = riS or else int S = $\emptyset$.

<u>Corollary 3</u>. Let S be a convex set. Then $C\ell\, S$ is also convex.

<u>Proof</u>: The result is trivial if S is empty. Let $x, y \in C\ell\, S$. We want to show that $\lambda x + (1 - \lambda)y \in C\ell\, S$ for each $\lambda \in (0, 1)$. Since riS is not empty by Theorem 2.3.4 then choose $z \in$ riS. Letting $x_\mu = \mu z + (1 - \mu)x$ it then follows by Theorem 2.3.5 above that $x_\mu \in$ riS for each $\mu \in (0, 1)$. Now $\lambda x_\mu + (1 - \lambda)y \in$ riS by Theorem 2.3.5 and hence $\lambda x_\mu + (1 - \lambda)y \in C\ell\, S$ for any fixed $\lambda \in (0, 1)$ and all $\mu \in (0, 1)$. Letting $\mu \to 0$ the result is at hand.

It should be noted that the result of Corollary 3 can be proved directly by definition of the closure and convexity of S. The purpose here, however, is to indicate how to use Theorem 2.3.4 and 2.3.5 to prove the corollary.

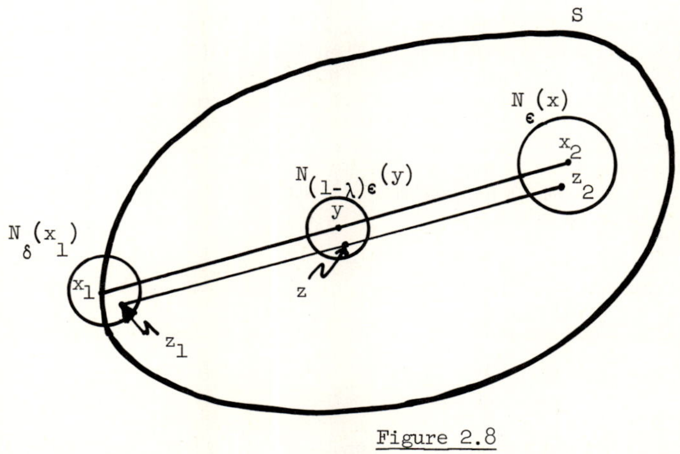

<u>Figure 2.8</u>

<u>Line Segment Between Point in the Closure and Point in the Relative Interior</u>

We will now give an important result which characterizes the relative interior points of convex sets. We will make use of the above results in doing so.

2.3.6 Theorem. Let S be a nonempty convex set in E_n. Then $x_o \in$ riS if and only if for each $x \in S$ there exists a $\mu > 1$ (possibly depending on x) such that $x_\mu = (1 - \mu)x + \mu x_o \in S$.

Proof: If $x_o \in$ riS then there is an $\epsilon > 0$ such that $N_\epsilon(x_o) \cap M(S) \subset S$. In particular for any $x \in S$, $x_\mu = (1 - \mu)x + \mu x_o \in M(S)$. If we choose $\mu = 1 + \frac{\epsilon}{2\|x_o - x\|}$ then $x_\mu \in N_\epsilon(x_o)$ and so $x_\mu \in S$. Conversely suppose that x_o satisfies the condition of the theorem. Since riS $\neq \emptyset$ by Theorem 2.3.4 then let $x \in$ riS. By hypothesis there exists $\mu > 1$ such that $x_\mu = (1 - \mu)x + \mu x_o \in S$. Therefore, we get $x_o = \frac{1}{\mu} x_\mu + \left(1 - \frac{1}{\mu}\right)x$. Since $\frac{1}{\mu} \in (0, 1)$ then by Theorem 2.3.5 $x_o \in$ riX and the proof is complete.

It should be noted that the only if part of the theorem holds without any convexity requirement on S. On the other hand the if part does not hold in general in the absence of convexity. For example, let S be the rational numbers between zero and one excluding the end points. Then any point in S satisfies the condition of the theorem, but on the other hand riS $= \emptyset$.

The following theorem shows that the relative interior of a convex set, the convex set itself, and the closure of it are "close" to each other. We need the following lemma first. No convexity assumption is needed here.

2.3.7 Lemma. Let S be an arbitrary set in E_n. Then $M(S) = M(C\ell\ S)$.

Proof: If S is empty then $M(S) = M(C\ell\ S) = \emptyset$. Now by Lemma 1.2.6 any linear subspace is closed. Since M(S) is the translation of a linear subspace then it is closed. Therefore, $S \subset C\ell\ S \subset M(S)$ and so $M(S) \subset M(C\ell\ S) \subset M(M(S)) = M(S)$. This shows that $M(S) = M(C\ell\ S)$.

The reader may be tempted to believe that M(S) = M(riS). However, this is not true without convexity of S. As a matter of fact we will make use of Theorem 2.3.8 below to show that M(S) = M(riS).

2.3.8 Theorem. Let S be a convex set in E_n. Then $C\ell$ (riS) $= C\ell\ S$ and ri$(C\ell\ S) =$ riS.

Proof: The result is trivial if S is empty. $riS \subset S$ implies that $C\ell(riS) \subset C\ell\, S$. On the other hand let $y \in C\ell\, S$ and choose $x \in riS$ (such an x exists by Theorem 2.3.4). Consider $\lambda x + (1 - \lambda)y$. By Theorem 2.3.5 $\lambda x + (1 - \lambda)y \in riS$ for each $\lambda \in (0, 1)$ and hence $y = \lim_{\lambda \to 0} \lambda x + (1 - \lambda)y \in C\ell(riS)$. This shows that $C\ell(riS) = C\ell\, S$. Now $X \subset C\ell\, S$ and by Lemma 2.3.7 $M(S) = M(C\ell\, S)$ and hence by Lemma 2.3.3 $riS \subset ri(C\ell\, S)$. To show the converse set inclusion let $y \in ri(C\ell\, S)$ and let $x \in riS$ (such x exists by Theorem 2.3.4). We will show that $y \in riS$. Now if $x = y$ the result is immediate, so we may as well assume that $x \neq y$. Since $y \in ri(C\ell\, S)$ there is an $\epsilon > 0$ such that $N_\epsilon(y) \cap M(C\ell\, S) \subset C\ell\, S$. Let $z = \left(1 + \frac{\epsilon}{2\|x-y\|}\right)y - \frac{\epsilon}{2\|x-y\|}\, x$. We claim that $z \in C\ell\, S$. Note that $y, x \in C\ell\, S$, $1 + \frac{\epsilon}{2\|x-y\|} - \frac{\epsilon}{2\|x-y\|} = 1$ which imply that $z \in M(C\ell\, S)$. Furthermore $\|z - y\| = \epsilon/2$ and hence $z \in C\ell\, S$. Rearranging the terms in the definition of z we get $y = \lambda z + (1 - \lambda)x$ where $\lambda = \dfrac{1}{1 + \frac{\epsilon}{2\|x-y\|}} \in (0, 1)$. By Theorem 2.3.5 it is clear that $y \in riS$ and the proof is complete.

It may be noted that the above theorem is false in the absence of convexity. For example, let S be the set of rational numbers between 0 and 1. It is obvious that $riS = \emptyset$ whereas $C\ell\, S = [0, 1]$. Therefore $C\ell(riS) = \emptyset$ and meanwhile $C\ell\, S = [0, 1]$. Also $ri(C\ell\, S) = (0, 1)$ but $riS = \emptyset$.

2.3.9 Lemma. Let S be a convex set in E_n. Then $M(riS) = M(S) = M(C\ell\, S)$.

Proof: In view of Lemma 2.3.7 we need to show that $M(riS) = M(C\ell\, S)$. By Theorem 2.3.8 above we get $C\ell(riS) = C\ell\, S$. Combining this and Lemma 2.3.7 we get $M(riS) = M(C\ell(riS)) = M(C\ell\, S)$ and the proof is complete.

As mentioned earlier convexity is needed here. If S is the set of rational numbers between zero and one then $M(riS) = \emptyset$ but on the other hand $M(S) = M(C\ell\, S) = E_1$.

The following theorem shows that $ri(S_1 + S_2) = riS_1 + riS_2$ when S_1 and S_2 are convex sets. This result will be used in different contexts. Actually we will use this fact to get some important separation theorems between disjoint convex sets.

2.3.10 Theorem. Let S_1 and S_2 be convex sets in E_n. Then $ri(S_1 + S_2) = riS_1 + riS_2$.

Proof: First note that $C\ell\, X_1 + C\ell\, X_2 \subset C\ell(X_1 + X_2)$. Note that convexity of X_1 and X_2 is not required here. At any rate let $X_1 = riS_1$ and $X_2 = riS_2$. Therefore

$C\ell(riS_1 + riS_2) \supset C\ell(riS_1) + C\ell(riS_2) = C\ell\, S_1 + C\ell\, S_2$ by Theorem 2.3.8. Denoting $riS_1 + riS_2$ by S we get $C\ell\, S \supset S_1 + S_2 \supset S$ and so $M(C\ell\, S) \supset M(S_1 + S_2) \supset M(S)$. But by Lemma 2.3.7 we have $M(C\ell\, S) = M(S)$ and so $M(C\ell\, S) = M(S_1 + S_2)$. By Lemma 2.3.3 it follows that $ri(C\ell\, S) \supset ri(S_1 + S_2)$. But again by Theorem 2.3.8 $ri(C\ell\, S) = riS = ri(riS_1 + riS_2)$. Therefore $ri(riS_1 + riS_2) \supset ri(S_1 + S_2)$ and hence $riS_1 + riS_2 \supset ri(S_1 + S_2)$. To show the converse let $x_o \in riS_1 + riS_2$. We will show that $x_o \in ri(S_1 + S_2)$. Since $x_o \in riS_1 + riS_2$ then $x_o = x_1 + x_2$ with $x_1 \in riS_1$ and $x_2 \in riS_2$. Now let z be an arbitrary element of $S_1 + S_2$, i.e., $z = z_1 + z_2$ with $z_1 \in S_1$, $z_2 \in S_2$. Now consider the set (S_1, S_2) which is convex. $(x_1, x_2) \in (riS_1, riS_2) = ri(S_1, S_2)$ and so by Theorem 2.3.6 for each $(z_1, z_2) \in (S_1, S_2)$ there must exist $\mu > 1$ such that $(1 - \mu)(z_1, z_2) + \mu(x_1, x_2) \in (S_1, S_2)$, i.e., $(1 - \mu)z_1 + \mu x_1 \in S_1$ and $(1 - \mu)z_2 + \mu x_2 \in S_2$. Adding we get $(1 - \mu)(z_1 + z_2) + \mu x_o \in S_1 + S_2$, i.e., $(1 - \mu)z + \mu x_o \in S_1 + S_2$. By Theorem 2.3.6 it follows that $x_o \in ri(S_1 + S_2)$ and the proof is complete.

Corollary. Let S_1 and S_2 be convex sets in E_n. Then $ri(S_1 - S_2) = riS_1 - riS_2$.

Proof: Note that $riS = -ri(-S)$. This can be immediately checked by noting that $M(S) = -M(-S)$. Now by Theorem 2.3.10 above we get $ri(S_1 - S_2) = ri(S_1 + (-S_2)) = riS_1 + ri(-S_2) = riS_1 - riS_2$.

It should be noted that the above theorem does not hold in general if S_1 and S_2 are not convex. For example let S_1 be the set of rational numbers between 0 and 1 and S_2 be the set of irrational numbers between 0 and 1 after adjoining the end points 0 and 1. Then $riS_1 = riS_2 = \emptyset$. On the other hand $S_1 + S_2 = [0, 2]$ and $ri(S_1 + S_2) = (0, 2)$. This shows that $riS_1 + riS_2 \not\supset ri(S_1 + S_2)$. As a matter of fact the converse relationship $riS_1 + riS_2 \subset ri(S_1 + S_2)$ does not hold in general in the absence of convexity. For example let S_1 be the set of irrational numbers between 0 and 1 and let S_2 be the interval [0, 1]. Then $S_1 + S_2$ is the set of irrational numbers between 0 and 2 and $ri(S_1 + S_2) = \emptyset$. On the other hand $riS_1 = \emptyset$ and $riS_2 = (0, 1)$. Hence $riS_1 + riS_2 \not\subset ri(S_1 + S_2)$. This shows the necessity of convexity assumption in the above theorem.

We will now talk about the notion of a boundary and relative boundary of a set.

2.3.11 Definition. The boundary of set S, denoted by ∂S is the set of points in

$C\ell$ S but not in int S, i.e., $\partial S = C\ell\ S \sim \text{int}\ S$. The relative boundary of S, denoted by rbS is the set of points in $C\ell$ S but not in the relative interior of S, i.e., $\text{rbS} = C\ell\ S \sim \text{riS}$.

From the above it is clear that every boundary point is also a relative boundary point. The reader may also note that every extreme point is indeed a relative boundary point. For example, consider the set $S = \{(x, y, 0) : 0 \le x \le 1, 0 \le y \le 1\}$. It can be immediately checked that the boundary of S is all S, the relative boundary of S is the set $\{(x, y, 0) : x = 1, 0 \le y \le 1\} \cup \{x, y, 0) : x = 0, 0 \le y \le 1\} \cup \{(x, y, 0) : 0 \le x \le 1, y = 1\} \cup \{(x, y, 0) : 0 \le x \le 1, y = 0\}$. Finally the extreme points are (0, 0, 0), (0, 1, 0), (1, 0, 0), and (1, 1, 0).

From the above definition it is clear that $y \in \partial S$ if $y \in C\ell\ S$ but $y \notin \text{int}\ S$. This means that any ball about y will contain point in S and points outside S. This means that we can find a sequence $\{y_k\}$ outside S which converges to y. On the other hand if $y \in \text{rbS}$ then we can find a sequence in M(S) outside S which converges to y.

Actually making use of the fact that for a convex set $\text{ri}(C\ell\ S) = \text{riS}$ and $\text{int}\ (C\ell\ S) = \text{int}\ S$, one can strengthen the above statements about points in the boundary and relative boundary of a convex set. As a matter of fact, for a convex set S, if $y \in \partial S$ then we can find a sequence $\{y_k\}$ which converges to S and meanwhile $y_k \notin C\ell\ S$ for each k. We will make use of this fact in proving Theorem 2.4.7. Similarly for a convex set S, if $y \in \text{rbS}$ then we can find a sequence $\{y_k\}$ converging to y with $y_k \in M(S)$ and meanwhile $y_k \notin C\ell\ S$ for each k.

The following result shows that a nontrivial bounded convex set has a nonempty relative boundary. By a nontrivial convex set we mean a set which is neither empty nor singleton. The result also shows that for the mentioned set any point in the relative interior is a convex combination of points in the relative boundary.

2.3.12 Lemma

Let S be a nonempty nonsingleton bounded convex set in E_n. Then $\text{rbS} \neq \emptyset$ and $\text{riS} \subset H(\text{rbS})$.

Proof: We will first construct a point in the relative boundary of S. Let $x_o \in \text{riS}$ and let $x \in M(S)$ but $x \notin C\ell\ S$. Such an x exists since S is bounded and non-singleton. Consider the ray R joining the points x and x_o, i.e., $R = \{\mu x + (1 - \mu)x_o$

$: \mu \in E_1\}$. Consider the function f defined on R by $f(y) = \|y - x_o\|$. f is continuous and hence achieves a maximum over the nonempty compact set $R \cap C\ell\, S$, say at y_o. In other words $y \in C\ell\, S \cap R$ implies that $\|y_o - x_o\| \geq \|y - x_o\|$. Now consider the sequence $\{\mu y_o + (1 - \mu)x_o\}$ in R where $\mu > 1$. $\|\mu y_o + (1 - \mu)x_o - x_o\| = \mu\|y_o - x_o\| > \|y_o - x_o\|$ for $\mu > 1$. This shows that $\mu y_o + (1 - \mu)x_o \notin C\ell\, S$ for each $\mu > 1$. By letting $\mu \to 1^+$ we generate a sequence $\{\mu y_o + (1 - \mu)x_o\}$ which belongs to R and meanwhile belongs to the complement of S. This means that $y_o \notin$ riS and since $y_o \in C\ell\, S$ then $y_o \in$ rbS by definition of the latter. This shows that rbS is not empty. To show that any point in the relative interior of S belongs to the convex hull of the relative boundary of S let x_o be a point in riS and construct $y_o \in$ rbS as above.

Now consider points of the form $\mu x_o + (1 - \mu)y_o$ with $\mu > 1$. Let $g(\mu) = \|(\mu - 1)(x_o - y_o)\|$ and maximize g over $\mu \geq 1$ satisfying $\mu x_o + (1 - \mu)y_o \in S$. g is continuous and achieves a maximum over the mentioned compact set at a point $z_o = \mu_o x_o + (1 - \mu_o)y_o$ with $\mu_o > 1$. As before it can be easily shown that $z_o \in$ rbS. Therefore $x_o = \frac{1}{\mu_o} z_o + \left(1 - \frac{1}{\mu_o}\right)y_o$ with $\frac{1}{\mu_o} \in (0, 1)$. In other words $x_o \in$ riS can be written as a convex combination of y_o and z_o in the relative boundary of S. This completes the proof.

Needless to say that the boundedness assumption is crucial here. For example any affine manifold has an empty relative boundary since the relative interior of an affine manifold is the manifold itself. Convexity is not crucial as long as the closure of the set under consideration has a point in its relative interior.

2.4 Support and Separation Theorems

Support and separation theorems play an extremely important role in the area of nonlinear programming. The statement that a point is an optimal solution of a nonlinear programming problem can be transformed into disjointness of some sets. Optimality criteria can then be developed by appealing to separation theorems of convex sets with empty intersection.

In this section we will develop some important support and separation theorems for disjoint convex sets. First consider the following definition of a hyperplane.

2.4.1 Definition. H is said to be a hyperplane (passing through x_o) if $H = \{x : \langle x - x_o, p\rangle = 0\}$ for some nonzero $p \in E_n$. p is usually called the normal vector of H.

Letting $\langle x_o, p\rangle = \alpha$, clearly H can be expressed as $H = \{x : \langle x, p\rangle = \alpha\}$. It is obvious that H is an affine manifold and the linear subspace L parallel to it is given by $L = \{x : \langle x, p\rangle = 0\}$. We have shown in Section 1.1.3 that dim $H = n - 1$. Also note that p is in the orthogonal complement of L, $L^{\perp}$. If a point x_o is specified then by definition p is orthogonal to all points $x - x_o$ where $x \in H$.

Indeed given any hyperplane H, one can define two sets in a natural way, namely $H^- = \{x : \langle x, p\rangle \le \alpha\}$ and $H^+ = \{x : \langle x, p\rangle \ge \alpha\}$.

These two sets are called halfspaces. It is obvious that $H^- \cup H^+ = E_n$. It can be easily checked that the two halfspaces above are indeed closed sets. One can define open halfspaces by changing the inequalities to strict inequalities in H^- and H^+, i.e., $\{x : \langle x, p\rangle < \alpha\}$ and $\{x : \langle x, p\rangle > \alpha\}$ define open halfspaces.

Now consider the following definition of a supporting hyperplane of a set.

2.4.2 Definition. Let S be a subset of E_n. A hyperplane H is called a supporting hyperplane of S if $S \subset H^-$ (or $S \subset H^+$) and $C\ell\, S \cap H \neq \emptyset$. H is called a proper supporting hyperplane of S if in addition $S \cap H \neq S$.

From the above definition we have to have two requirements if H to support a set S. The first of these is that the set should be completely contained in one of the halfspaces generated by H. The second requirement is that the closure of S and H should have at least one point in common. If $x_o \in C\ell\, S \cap H$ then we say that H supports the set S at x_o. It is obvious that a hyperplane H may support a set S at several distinct points.

Figure 2.9 illustrates 3 cases of supporting hyperplanes. In Figure 2.9(a) the supporting hyperplane to S at x_o is unique. Obviously, in general, a supporting hyperplane to S at a given point x_o need not be unique, as in Figure 2.9(b). This mainly depends on whether or not the set S is "smooth" at the point x_o. This notion of smoothness will be further explored in the next Chapter. Finally Figure 2.9(c) shows an improper supporting hyperplane since the set S is completely contained in H. This type of support is of no primary interest.

The following lemma proves the intuitively obvious fact that $\inf_{x \in S} \langle x, p\rangle$ is indeed α if the hyperplane $H = \{x : \langle x, p\rangle = \alpha\}$ supports S and S is contained in H^+.

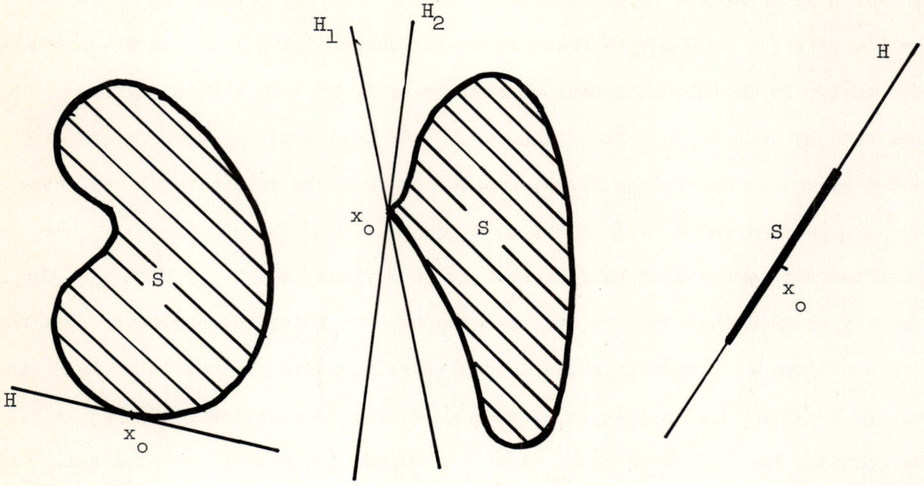

Figure 2.9

Examples of Supporting Hyperplanes

2.4.3 Lemma. Let $H = \{x : \langle x, p \rangle = \alpha\}$ be a supporting hyperplane of S such that $S \subset H^+$. Then $\inf \{\langle x, p \rangle : x \in S\} = \alpha$.

Proof: Since $S \subset H^+$ then $\langle x, p\rangle \geq \alpha$ for each $x \in S$, i.e., $\inf \{\langle x, p\rangle : x \in S\} \geq \alpha$. If equality does not hold then there must exist an $\epsilon > 0$ with $\langle x, p\rangle \geq \alpha + \epsilon$ for each $x \in S$. But by definition of a supporting hyperplane $C\ell\ S \cap H \neq \emptyset$, i.e., there exists $y \in C\ell\ S$ with $\langle y, p\rangle = \alpha$. But $y \in C\ell\ S$ implies that there exists a sequence $\{x_n\}$ in S which converges to y and hence $\langle y, p \rangle = \lim_{n\to\infty} \langle x_n, p\rangle \geq \alpha + \epsilon$, a contradiction. This completes the proof.

The following theorem is a projection theorem on closed convex sets. It represents an extension of the projection theorem on linear subspaces presented in Chapter 1. Here we wish to find the minimum distance between a given point and a closed convex set. The theorem says that there is a unique minimizing point and it also gives a characterization of this point. The proof of the theorem is similar to that of

Theorem 1.3.3 and hence the proof is omitted. As a matter of fact proving the existence of the unique minimizing point is identical to that of Theorem 1.3.3 and the proof of the characterization of the minimizing point is similar but not identical to that of the mentioned theorem. Figure 2.10 below shows the unique minimizing point corresponding to a closed convex and a point outside the set.

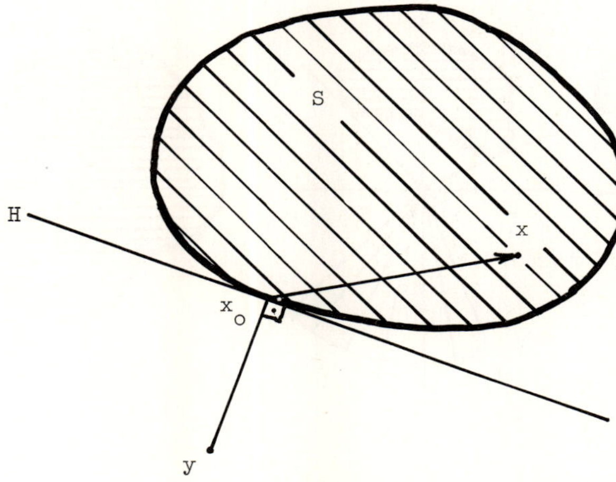

Figure 2.10

Projection of a Point on a Closed Convex Set

2.4.4 Theorem. Let S be a closed convex set in E_n. Let y be a point in E_n with $y \notin S$. Then there is a unique $x_o \in S$ such that $\|y - x_o\| = \min \{\|y - x\| : x \in S\}$. Furthermore x_o is such a minimizing point if and only if $\langle x - x_o, y - x_o \rangle \leq 0$ for each $x \in S$.

At this stage it may be helpful to introduce the definition of the polar of a set. From this we will get a generalization of the decomposition on a linear subspace and its orthogonal complement of Chapter 1. We will show that given a closed convex cone and its polar, any point in E_n can be uniquely represented as the sum of two points in the cone and its polar. First consider the following definition of a polar.

2.4.5 Definition. Let S be an arbitrary set in E_n. The polar of S, denoted by S^* is the set of all ζ such that $\langle x, \zeta \rangle \leq 0$ for each $x \in S$. If S is empty we will interpret S^* as E_n.

From the above definition it is clear that the polar of a set S is a convex cone since ζ_1 and $\zeta_2 \in S^*$ imply that $\lambda\zeta_1 + \mu\zeta_2 \in S^*$ for each $\lambda, \mu \geq 0$. Also S^* is closed since any convergent sequence in S^* will converge to a point in S^*. Figure 2.11 below gives some examples of polar cones.

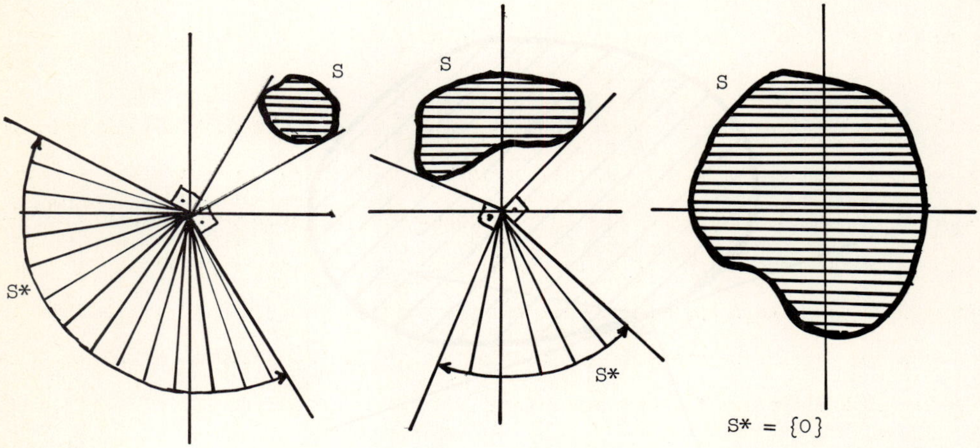

Figure 2.11

Some Examples of Polar Cones

As an important example, suppose that L is a linear subspace in E_n. It is clear that the polar of L is indeed its orthogonal complement. This is shown as follows. Let $\zeta \in L^*$, i.e., $\langle x, \zeta \rangle \leq 0$ for each $x \in L$. But since L is a subspace then $-x \in L$ and so we get $\langle -x, \zeta \rangle \leq 0$. This shows that $\langle x, \zeta \rangle = 0$ for each $x \in L$ and hence $\zeta \in L^\perp$, i.e., $L^* \subset L^\perp$. The converse is trivially checked and hence $L^* = L^\perp$.

Now suppose that we have a closed convex cone C. By Theorem 2.4.4 given any point y there is a unique point $x_o \in C$ such that $\|y - x_o\| = \min\{\|y - x\| : x \in C\}$. Further we have the characterization $\langle x - x_o, y - x_o \rangle \leq 0$ for each $x \in C$. But since C is a cone then given any point z in C we must have $x_o + z \in C$ and we get $\langle z, y - x_o \rangle \leq 0$ for each $z \in C$. But this immediately shows that $y - x_o \in C^*$. In essence then we have shown that given a point y we can decompose it into a point x_o in C and $y - x_o$ in C^*. In other words, we have shown that E_n is thesum of C and C^*. In Figure 2.12 (i), C is a half-space and C^* is the normal vector. It is obvious that y can be decomposed into x_o and $y - x_o$. One can ask the question whether, in general such a decomposition is unique. The answer is no, i.e., we can find different decompositions (representations) of the same point. In Figure 2.12 (ii) y can be decomposed into x_o and $y - x_o$, and also into x_1 and $y - x_1$ where x_o and x_1 are in C, and $y - x_o$ and $y - x_1$ are in C^*.

We have seen how to make use of Theorem 2.4.4 to get a generalized decomposition result in closed convex cones and their polars. We will now come back to the main issue of this section which is support and separating theorems of convex sets. The following remark shows that if we have a closed convex set and a point outside the set then we can find a hyperplane that the convex set lies in one of the halfspaces generated by the hyperplane and the other point lies in the interior of the other halfspace. This result follows easily from Theorem 2.4.4 above.

2.4.6 Lemma. Let S be a nonempty closed convex set in E_n and $y \notin S$. Then there exists a nonzero vector p in E_n such that $\inf\{\langle x, p\rangle : x \in S\} > \langle y, p\rangle$.

Proof: By Theorem 2.4.4 there is a unique $x_o \in S$ with $\langle x - x_o, y - x_o\rangle \leq 0$ for each $x \in S$. Note that $y \neq x_o$ and so $0 < \|y - x_o\|^2 = \langle y - x_o, y - x_o\rangle =$ $\langle y, y - x_o\rangle - \langle x_o, y - x_o\rangle$. Combining the above two inequalities we get $\langle x, y - x_o\rangle \leq \langle x_o, y - x_o\rangle < \langle y, y - x_o\rangle$ for each $x \in S$. Letting $x_o - y = p$ it follows that $\langle x, p\rangle \geq \langle x_o, p\rangle > \langle y, p\rangle$ for each $x \in S$ and the result follows.

Note that the above remark established a hyperplane $H = \{x : \langle x - x_o, p\rangle = 0\}$ with $S \subset H^+$ and $y \in \operatorname{int} H^-$. This is also clear by reviewing Figure 2.10.

The following theorem is a consequence of the above result. Here we relax the requirement that $y \notin S$ to the requirement that $y \in \partial S$, the boundary of S. Actually we will construct a hyperplane that passes through y and supports S. This is an important property of convex sets.

2.4.7 Theorem. Let S be a nonempty convex set in E_n and let $y \in \partial S$. Then there exists a nonzero $p \in E_n$ such that $\langle y, p\rangle = \inf\{\langle x, p\rangle \geq 0 : x \in S\}$, i.e., there is a hyperplane $H = \{x : \langle x - y, p\rangle = 0\}$ which supports S at y.

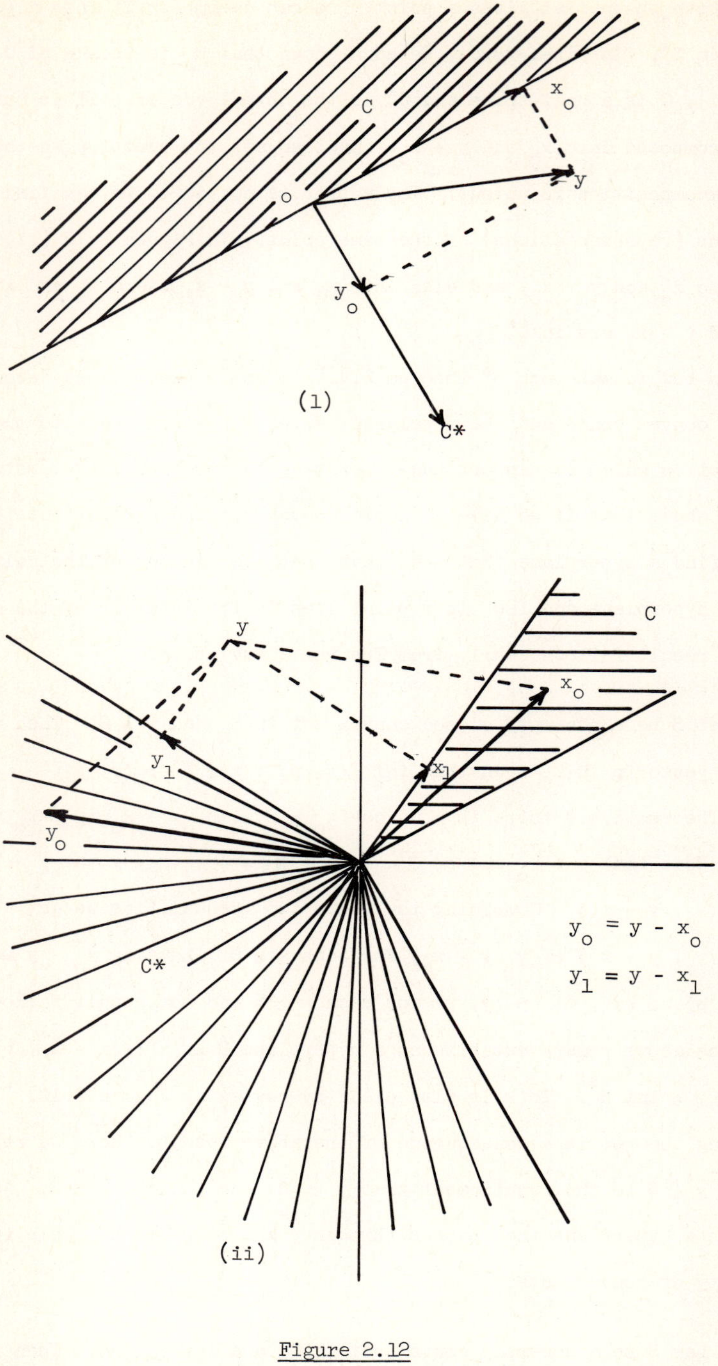

Figure 2.12

Decomposition of Points in E_n onto a Closed Convex Cone and its Polar

Proof: Since $y \in \partial S$ then there is a sequence $\{y_k\}$ outside $C\ell S$ such that $y_k \to y$ (see the discussion following Definition 2.3.11). By Lemma 2.4.6 for any $y_k \notin C\ell\, S$ there exists a nonzero p_k such that $\langle x, p_k\rangle > \langle y_k, p_k\rangle$ for each $x \in C\ell\, S$. Since $p_k \neq 0$ then we get $\langle x, \frac{p_k}{\|p_k\|}\rangle < \langle y_k, \frac{p_k}{\|p_k\|}\rangle$ for each $x \in C\ell\, S$. $\{\frac{p_k}{\|p_k\|}\}$ is a sequence of norm 1 and hence has a convergent subsequence $\frac{p_{k_j}}{\|p_{k_j}\|} \to p$ with $\|p\| = 1$. Considering this subsequence we get $\langle x, \frac{p_{k_j}}{\|p_{k_j}\|}\rangle > \langle y_{k_j}, \frac{p_{k_j}}{\| p_{k_j}\|}\rangle$ and as we take the limit as $j \to \infty$ we get $\langle x, p\rangle \geq \langle y, p\rangle$ for each $x \in C\ell\, S$. In other words for any $x \in C\ell\, S$ we have $\langle x - y, p\rangle \geq 0$ and the proof is complete.

Now consider the set $S = \{(x, y) : 0 \leq x \leq 2,\ x + y = 1\}$. Can we find a supporting hyperplane of S at the point (1, 0)? According to the above theorem this is possible and it can be easily checked that the only supporting hyperplane is $H = \{(x, y) : x + y = 1\}$. Clearly this supporting hyperplane is not proper since $S \subset H$ and such a hyperplane is not of primary interest in many applications. We will then define Theorem 2.4.7 above to show that every nonempty convex set has a proper supporting hyperplane at every point in the relative boundary. We will now develop the following lemma which is needed to prove the desired theorem.

2.4.8 Lemma. Let S be a nonempty convex set in E_n. Let $H = \{x : \langle x - y, p\rangle = 0\}$ be a supporting hyperplane of S at $y \in C\ell\, S$, where p is a nonzero vector in E_n. Suppose that $p \in L$ where L is the subspace parallel to M(S), i.e., $M(S) = y + L$. Then $H \cap \mathrm{ri}S = \emptyset$. In particular H is a proper supporting hyperplane of S.

Proof: Since H supports S at y then we may assume that $S \subset H^+$. Suppose to the contrary of the desired result that there is a $x \in H \cap \mathrm{ri}S$. Therefore $\langle x - y, p\rangle = 0$ and there is an $\epsilon > 0$ such that $N_\epsilon(x) \cap M(S) \subset S$. Since $p \in L$ then $y + p \in M(S)$ and therefore for $\lambda < 0$ sufficiently small we have $\lambda(y + p) + (1 - \lambda)x \in S$. But $S \subset H^+$ implies that $\langle\lambda(y + p) + (1 - \lambda)x - y, p\rangle \geq 0$, i.e., $\langle(1 - \lambda)(x - y), p\rangle + \lambda\|p\|^2 \geq 0$. But since $\langle x - y, p\rangle = 0$ we get $\lambda\|p\|^2 \geq 0$ which is a contradiction since $\lambda < 0$ and $p \neq 0$. This completes the proof.

We conclude from the above lemma that H properly supports S at y. It is then clear that $\inf\{\langle x - y, p\rangle : x \in S\} \geq 0$ and $\sup\{\langle x - y, p\rangle : x \in S\} > 0$ because otherwise we violate the conclusion that S is not contained in H.

2.4.9 <u>Theorem</u>. Let S be a nonempty convex set in E_n and let $y \in$ rbS. Then there exists a hyperplane H that supports S at y in such a way that $H \cap \text{ri}S = \emptyset$, i.e., the particular H supports S properly.

<u>Proof</u>: Since $y \in$ rbS then there is a sequence $\{y_k\}$ in M(S) which converges to y and such that $y_k \notin C\ell\, S$ for each k (see the discussion following Definition 2.3.11). In view of Lemma 2.4.6 we must have a nonzero p_k such that $\inf\{\langle x, p_k\rangle : x \in S\} > \langle y_k, p_k\rangle$ where $p_k = \dfrac{y_k - x_k}{\|y_k - x_k\|}$ and x_k is the unique minimizing point corresponding to y_k for each $k \geq 1$. Since $y_k \in M(S)$ and $x_k \in C\ell\, S$ then $y_k - x_k \in L$, the subspace parallel to M(S) and so is $p_k = \dfrac{y_k - x_k}{\|y_k - x_k\|}$. $\{p_k\}$ is a bounded sequence in L and must have a convergent subsequence $\{p_{k_j}\}$ with limit p in L since L is closed. Considering this subsequence we get $\langle x - y_{k_j}, p_{k_j}\rangle > 0$ for each $x \in S$. Letting $j \to \infty$ we have $\langle x - y, p\rangle \geq 0$ for each $x \in S$, i.e., $H = \{x : \langle x - y, p\rangle = 0\}$ supports S at y. But since $p \in L$ the result is immediate by Lemma 2.4.8.

Now consider the set of points in E_3 which satisfy the following system of inequalities: $x + y - z \leq 2$, $-2x + y + 3z \leq 4$, $x \geq 0$, $y \geq 0$, $z \geq 0$. Clearly this set is closed and convex and the point (2, 2, 2) is a boundary point of this set. Now how can we find a supporting hyperplane of this set at the point (2, 2, 2)? First note that the points (2, 2, 2) lies in the intersection of the two hyperplanes $\{(x, y, z) : x + y - z = 2\}$ and $\{(x, y, z) : -2x + y + 3z = 4\}$. It is also clear that the set of interest is a subset of each of the halfspaces $\{(x, y, z) : x + y - z \leq 2\}$ and $\{(x, y, z) : -2x + y + 3z \leq 4\}$, i.e., either of the hyperplanes $\{(x, y, z) : x + y - z = 2\}$ and $\{(x, y, z) : -2x + y + 3z = 4$ is a supporting hyperplane to the set at the point (2, 2, 2). As a matter of fact, any convex combination of these is indeed a supporting hyperplane of the set at the point (2, 2, 2). In other words, any hyperplane passing through (2, 2, 2) with normal vector $\lambda(1, 1, -1) + (1 - \lambda)(-2, 1, 3)$ where $\lambda \in [0, 1]$ is a supporting hyperplane of the set at (2, 2, 2). So we get $H = \{(x, y, z) : \langle(-2 + 3\lambda, 1, 3 - 4\lambda), (x - 2, y - 2, z - 2)\rangle = 0\}$, i.e., $H = \{(x, y, z) : (-2 + 3\lambda)x + y + (3 - 4\lambda)z = 4 - 2\lambda\}$ will support the set for any $\lambda \in [0,1]$.

The following theorem is useful in developing different results later. We assert that there is no proper supporting hyperplane to a convex set at a point in the relative interior.

2.4.10 Theorem. Let S be a nonempty convex set in E_n and let $y \in \mathrm{ri}S$. Then there is no proper supporting hyperplane of S passing through y.

Proof: Suppose to the contrary of the desired result that such a proper supporting hyperplane H exists, i.e., $S \subset H^+$, where $H = \{x : \langle x - y, p\rangle = 0\}$. Now let $x \in S$ be such that $\langle x - y, p\rangle \neq 0$ (such an x exists since $S \cap H \neq S$). Consider points of the form $\alpha x + (1 - \alpha)y \in M(S)$. Since $y \in \mathrm{ri}S$ then for α sufficiently small $x_\alpha = \alpha x + (1 - \alpha)y \in S$, i.e., there is an $\bar{\alpha} > 0$ such that $x_\alpha \in S$ for each $\alpha \in [-\bar{\alpha}, \bar{\alpha}]$. Therefore, for $\alpha = -\bar{\alpha} \dfrac{\langle x - y, p\rangle}{|\langle x - y, p\rangle|}$, $x_\alpha \in S$. For this α we get $\langle x_\alpha - y, p\rangle$ $= -\alpha\langle x - y, p\rangle = -\bar{\alpha}|\langle x - y, p\rangle| < 0$ contradicting the hypothesis that $S \subset H^+$. Hence no proper supporting hyperplane exists and the proof is complete.

Corollary. Let S be a convex set in E_n with nonempty interior and let $y \in \mathrm{int}\, S$. Then there is no supporting hyperplane of S through y.

2.4.11 Theorem. Let S be a nonempty compact convex set in E_n and let E be the set of extreme points of S. Then $E \neq \emptyset$ and $S = H(E)$.

Proof: We prove the above result by induction on the dimension of the set S. So let dim S = 0, i.e., S consists of a single point. Obviously the statement of the theorem holds. Now suppose that dim S = k and that the theorem is true for compact convex sets with dimension less than k. Let x_0 be a point in the relative boundary of S which is not empty by Lemma 2.3.12. Thus by Theorem 2.4.9 there is a supporting hyperplane $H = \{x : \langle x - x_0, p\rangle = 0\}$ such that $S \subset \{x : \langle x - x_0, p\rangle \geq 0\}$. Here p is a nonzero vector in E_n and $H \cap \mathrm{ri}S = \emptyset$. Note that $H \cap S$ is a compact convex set. Also note that $k = \dim S > \dim (H \cap S)$. This follows since the affine manifold generated by $H \cap S$ is a subset of H which does not intersect riS, which implies that $M(H \cap S)$ is a proper subset of $M(S)$. Therefore by the induction hypothesis the compact convex set $H \cap S$ has extreme points and the convex hull of these points is indeed $H \cap S$. We will now show that if e is an extreme point of $H \cap S$ then e is an extreme point of S. Now suppose that e is an extreme point of $H \cap S$ and suppose that $e = \lambda x + (1 - \lambda)y$ where $x, y \in S$ and $\lambda \in (0,1)$. Since $e \in H$ then $0 = \langle e - x_0, p\rangle = \langle \lambda x + (1-\lambda)y - x_0, p\rangle$. But since $x, y \in S$ then $\langle x - x_0, p\rangle \geq 0$ and $\langle y - x_0, p\rangle \geq 0$. This means that $\langle \lambda x + (1-\lambda)y - x_0, p\rangle = 0$ if and only if $\langle x - x_0, p\rangle = \langle y - x_0, p\rangle = 0$. In other words

$x,y \in H$. This implies that $x,y \in H \cap S$ and since e is an extreme point of $H \cap S$ then $x = y = e$. This shows that the set of extreme points E of S is not empty. Since x_o can be represented as a convex combination of the extreme points of $S \cap H$ then it can also be represented as a convex combination of the extreme points of S since the former are a subset of the latter. Since this is true for every relative boundary point then it is certainly true that $rbS \subset H(E)$. But note that every point in the relative interior of S can be represented as a convex combination of points in rbS (see Lemma 2.3.12) and hence $S \subset H(E)$. On the other hand $E \subset S$ and by convexity of S we must have $H(E) \subset S$ and so $H(E) = S$. This completes the proof.

The following theorem shows that the number of extreme points of a set defined by a finite system of linear inequalities is finite. Needless to say that in general this is not the case. For example the unit ball $\{x : \|x\| \leq 1\}$ has infinitely many extreme points, namely the set $\{x : \|x\| = 1\}$.

2.4.12 Theorem. Let $S = \{x : Ax \geq b\}$ and let E be the set of extreme points of S. Here A is an $m \times n$ matrix and b is a m vector. Then either $E = \emptyset$ or else E consists of a finite number of points.

Proof: Let $A^t = (a_1, a_2, \ldots, a_m)$ where each a_i is an $n \times 1$ vector. Let $M(x) = \{i : \langle x, a_i \rangle - b_i > 0\}$. We will first show that $x,y \in E$ with $M(x) = M(y)$ implies that $x = y$. We will then make use of this result to show that E is finite. So let $x,y \in E$ with $M(x) = M(y)$. First consider the case when $M(x) = M(y) = \emptyset$, i.e., $Ax = Ay = b$. Consider the point $z = 2x - y$. Clearly $z \in S$ since $Az = 2Ax - Ay = b$. Meanwhile $x = \frac{1}{2}z + \frac{1}{2}y$ which means that $x = y = z$ since x is an extreme point of S. Now suppose that $M(x) = M(y) \neq \emptyset$. Let $\theta = \min\left\{\frac{\langle x, a_i \rangle - b_i}{\langle y, a_i \rangle - b_i} : i \in M(x)\right\} > 0$. This means that $\langle x, a_i \rangle - b_i \geq \theta[\langle y, a_i \rangle - b_i]$ for each $i \in M(x)$. But for $i \notin M(x)$ we have $\langle x, a_i \rangle = \langle y, a_i \rangle = b_i$. Therefore we get $Ax - b \geq \theta(Ay - b)$. If $\theta \geq 1$ and since $Ay \geq b$ we get $Ax - b \geq Ay - b$, i.e., $A(x-y) \geq 0$. Letting $z = x + (x-y)$, then $Az = Ax + A(x-y) \geq b$. In other words $z \in S$ and by definition of z, $x = \frac{1}{2}y + \frac{1}{2}z$. Since x is an extreme point of S then $x = y = z$. Now if $\theta < 1$ consider $z = \frac{1}{1-\theta}(x-\theta y)$. Then $Az = \frac{1}{1-\theta}(Ax-\theta Ay) \geq b$ since $Ax - b \geq \theta(Ay - b)$. Therefore $z \in S$. Again by definition of z, $x = \theta y + (1-\theta)z$ which implies that $x = y = z$ by noting that $\theta \in (0,1)$ and $x \in E$. So far we have shown that $x,y \in E$ with $M(x) = M(y)$ implies that $x = y$.

This means that $x \neq y$ implies that $M(x) \neq M(y)$. Note that we have a finite number of sets of the form $M(x)$ for x varying over E, namely we have 2^m such sets corresponding to all possible subsets of $\{1,2,\ldots,m\}$. This shows that the number of elements in E is finite.

Corollary: Let $S = \{x : Ax \geq b\}$ be nonempty and bounded and let E be its extreme points. Then $E \neq \emptyset$ and has finitely many points and $S = H(E)$.

Proof: Follows immediately from the above two theorems.

So far we have developed some results on supports of convex sets at boundary and relative boundary points. We will now make use of these results to develop separation theorems between nonintersecting convex sets. Now consider the following definition of separation of convex sets.

2.4.13 Definition. Let S_1 and S_2 be arbitrary sets in E_n. A hyperplane H is said to separate S_1 and S_2 if $S_1 \subset H^+$ and $S_2 \subset H^-$ (or else $S_1 \subset H^-$ and $S_2 \subset H^+$). H is called a proper separating hyperplane if in addition $S_1 \cup S_2 \not\subset H$. H is said to strictly separate S_1 and S_2 if $S_1 \subset \text{int } H^-$ and $S_2 \subset \text{int } H^+$ (or alternatively $S_1 \subset \text{int } H^+$ and $S_2 \subset \text{int } H^-$). H is said to strongly separate S_1 and S_2 if there is a ball B of radius $\epsilon > 0$ such that $S_1 + B \subset H^-$ and $S^2 + B \subset H^+$ (or alternatively $S_1 + B \subset H^+$ and $S_2 + B \subset H^-$).

Figure 2.13 below gives different types of separation of two sets.

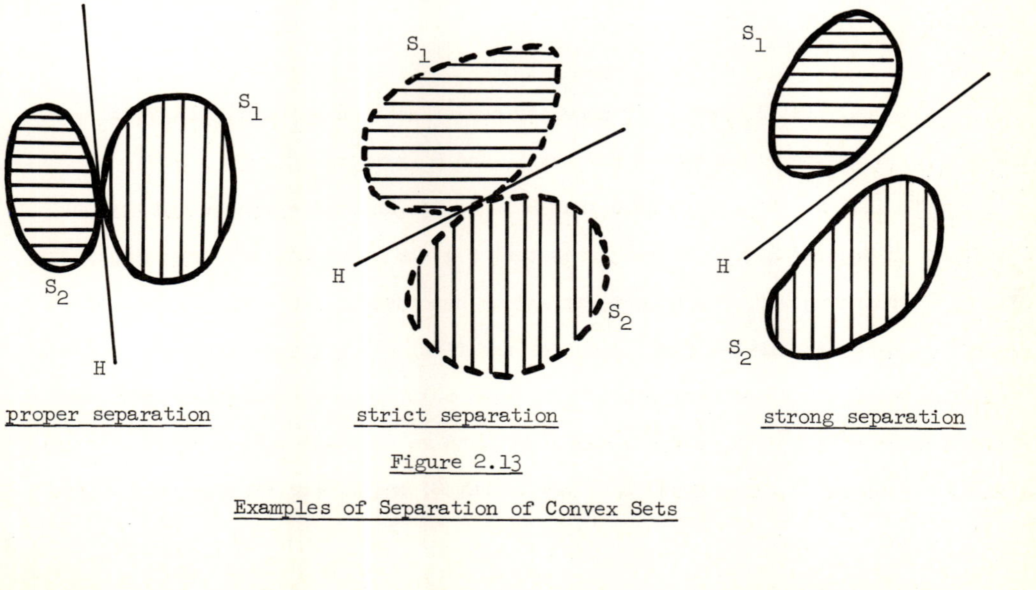

Figure 2.13

Examples of Separation of Convex Sets

From the above it is clear that strong separation implies strict separation. The latter indeed implies proper separation. For proper separation between S_1 and S_2 it is clear from the definition that there must exist a hyperplane $H = \{x : \langle x, p\rangle = \alpha\}$ with $S_1 \subset H^-$ and $S_2 \subset H^+$. Therefore, we get $\langle x, p\rangle \leq \alpha$ for each $x \in S_1$ and $\langle x, p\rangle \geq \alpha$ for each $x \in S_2$. This means that $\sup\{\langle x, p\rangle : x \in S_1\} \leq \alpha$ and $\inf\{\langle x, p\rangle : x \in S_2\} \geq \alpha$ and hence $\inf\{\langle x, p\rangle : x \in S_2\} \geq \sup\{\langle x, p\rangle : x \in S_1\}$. Furthermore since $S_1 \cup S_2 \not\subset H$ we must have either $\langle x, p\rangle < \alpha$ for some $x \in S_1$ or else $\langle x, p\rangle > \alpha$ for some $x \in S_2$. Hence we respectively get either $\inf\{\langle x, p\rangle : x \in S_1\} < \alpha$ or else $\sup\{\langle x, p\rangle : x \in S_2\} > \alpha$. At any rate it follows that $\sup\{\langle x, p\rangle : x \in S_2\} > \inf\{\langle x, p\rangle : x \in S_1\}$.

For strict separation between S_1 and S_2 there must exist a nonzero vector p and a real number α corresponding to the separating hyperplane such that $\langle x, p\rangle < \alpha$ for each $x \in S_1$ and $\langle x, p\rangle > \alpha$ for each $x \in S_2$. The reader may note that $\sup\{\langle x, p\rangle : x \in S_1\} \leq \inf\{\langle x, p\rangle : x \in S_2\}$. Strict inequality cannot be guaranteed as in Figure 2.13.

Finally strong separation between S_1 and S_2 amounts to finding a nonzero vector p such that $\inf\{\langle x, p\rangle : x \in S_2\} > \sup\{\langle x, p\rangle : x \in S_1\}$. As a matter of fact any real number $\alpha \in (\sup\{\langle x, p\rangle : x \in S_1\}, \inf\{\langle x, p\rangle : x \in S_1\})$ along with p corresponds to a strongly separating hyperplane, i.e., the hyperplane $H = \{x : \langle x, p\rangle = \alpha\}$ strongly separates S_1 and S_2 where α is any real number in the above open interval.

The following theorem shows that convex sets with no points common to their relative interiors can be properly separated.

2.4.14 <u>Theorem</u>. Let S_1 and S_2 be nonempty convex sets in E_n with $\text{ri}S_1 \cap \text{ri}S_2 = \emptyset$. Then there is a hyperplane that separates S_1 and S_2 properly.

<u>Proof</u>: Let $S = S_1 - S_2$. Note that S is convex and $\text{ri}S = \text{ri}S_1 - \text{ri}S_2$ by the corollary to Theorem 2.3.10. This means that $0 \notin \text{ri}S$ since $0 \in \text{ri}S$ would imply that there is a point in $\text{ri}S_1 \cap \text{ri}S_2$, violating our hypothesis. But $0 \notin \text{ri}S$ means that either $0 \notin C\ell\, S$ or else $0 \in \text{rb}S$. Consider the first case. By Lemma 2.4.6 there is a nonzero $p \in E_n$ such that $\inf\{\langle x, p\rangle : x \in S\} > \langle 0, p\rangle = 0$. But this means that $\inf\{\langle x_1, p\rangle - \langle x_2, p\rangle : x_1 \in S_1, x_2 \in S_2\} > 0$ and we conclude that $\inf\{\langle x, p\rangle : x \in S_1\} > \sup\{\langle x, p\rangle : x \in S_2\}$. This means that S_1 and S_2 are strongly separated

(and hence properly separated). Now consider the other case when $0 \in \mathrm{rb}S$. By Theorem 2.4.9 there is a hyperplane that supports S at the origin in such a way that $H \cap \mathrm{ri}S = \emptyset$. This means that there is a nonzero $p \in E_n$ such that $\inf\{\langle x, p\rangle : x \in S\} \geq \langle 0, p\rangle = 0$ and meanwhile $\sup\{\langle x, p\rangle : x \in S\} > 0$. This means that $\inf\{\langle x, p\rangle : x \in S_1\} \geq \sup\{x \in S_2\}$ and meanwhile $\sup\{\langle x, p\rangle : x \in S_1\} > \inf\{\langle x, p\rangle : x \in S_2\}$ and we get proper separation of S_1 and S_2. This completes the proof.

The following theorem is of the strong separation type. Here we need one of the sets be compact and the other set be closed.

2.4.15 Theorem. Let S_1 and S_2 be nonempty convex sets in E_n with $S_1 \cap S_2 = \emptyset$. If S_1 is compact (closed and bounded) and S_2 is closed then there is a hyperplane which strongly separates S_1 and S_2.

Proof: Let $S = S_1 - S_2$. We claim that S is closed. To show this let $\{z_k\}$ be a sequence in S converging to z, i.e., $x_k - y_k \to z$ with $x_k \in S_1$ and $y_k \in S_2$ for each k. We need to show that $z \in S$. Since S_1 is bounded then $\{x_k\}$ is bounded and hence has a convergent subsequence $x_{k_j} \to x$. But since S_1 is closed then $x \in S_1$. Now $x_{k_j} - y_{k_j} \to z$ and $x_{k_j} \to x$ imply that $y_{k_j} \to y$. Since S_2 is closed then $y \in S_2$. In other words we have shown that $z = x - y$ with $x \in S_1$ and $y \in S_2$, i.e., $z \in S$. Therefore, S is closed and furthermore $0 \notin S$ since $S_1 \cap S_2 = \emptyset$. Therefore, by Lemma 2.4.6 there is a nonzero $p \in E_n$ such that $\inf\{\langle z, p\rangle : z \in S\} > \langle 0, p\rangle = 0$. This implies that $\inf\{\langle x, p\rangle : x \in S_1\} > \sup\{\langle x, p\rangle : x \in S_2\}$, i.e., there is a hyperplane with normal p which strongly separates S_1 and S_2.

It should be noted that the above theorem does not hold if the boundedness assumption of S_1 is dropped. For example, let $S_1 = \{(x, y) : y \geq e^{-x}\}$ and $S_2 = \{(x, y) : y \leq e^{-x}\}$. It is clear that all the hypotheses of the above theorem hold except for boundedness of S_1. It is also clear that even though there is a hyperplane which strictly separates S_1 and S_2, one cannot find a hyperplane that strongly separates S_1 and S_2.

CHAPTER 3

CONVEX CONES

In Chapter 1 we discussed linear subspaces, i.e., a set having the property that it contains all linear combinations of points in the set. In Chapter 2 we considered the set containing all non-negative convex combinations of points in the set, namely a convex set. As seen earlier convex cones are sets whose definition is less restrictive than that of a convex set, but more restrictive than that of a subspace. Put in another way, the convex cone generated by a set contains the convex hull of the set but is contained in the subspace spanned by the set.

We mentioned earlier that the notion of convex sets and its properties are fundamental in the development of optimality conditions. These conditions can be restated in equivalent, but more usable form using the properties of the cones. One of the properties we will be developing is essentially a generalization of the Farkas lemma, a well-known and frequently used "theorem of the alternative" for linear systems.

3.1 Cones, Convex Cones, and Polar Cones

3.1.1 Definition. A set K in E_n is said to be a cone with vertex x_o if $x \in K$ implies that $x_o + \lambda(x - x_o) \in K$ for each $\lambda > 0$.

From now on, unless otherwise stated, the vertex of a cone is the origin, i.e., we will assume that $x_o = 0$. Further, the following may be worth noting:

(i) The vertex of a cone may or may not belong to the cone. However, the vertex belongs to the closure of the cone. Many authors, however, do include the vertex in the cone by letting $\lambda \geq 0$ in the definition of the cone.

(ii) A nontrivial cone K is an unbounded set, i.e., given any real number $M > 0$ we can find an $x \in K$ with $\|x\| > M$. By a nontrivial cone we mean a cone which is neither empty nor the zero vector.

(iii) A cone K may or may not be convex.

(iv) A cone K may be open, closed, or neither open nor closed.

3.1.2 Definition. A cone K is said to be pointed if whenever $x \neq 0$ is in the cone, then $-x$ is not in the cone.

It should be noted that pointed cones are contained in a closed half-space.

3.1.3 Definition. A set C is called a convex cone if it is a cone and also a convex set.

The following lemma is frequently used to characterize convex cones and is an immediate consequence of the above definition.

3.1.4 Lemma. C is a convex cone if and only if the following hold:

(i) $x \in C$ implies that $\lambda x \in C$ for each $\lambda > 0$.

(ii) $x_1, x_2 \in C$ implies that $x_1 + x_2 \in C$.

Proof: Suppose that C is a convex cone, then $x_1, x_2 \in C$ implies that $\lambda x_1 + (1 - \lambda)x_2 \in C$ for each $\lambda \in (0,1)$. Letting $\lambda = \frac{1}{2}$ we get $\frac{1}{2}x_1 + \frac{1}{2}x_2 \in C$ and hence $x_1 + x_2 \in C$. Conversely assume (i) and (ii). If $x_1, x_2 \in C$ then from (i) we get $\lambda x_1 \in C$ and $(1 - \lambda)x_2 \in C$ for each $\lambda \in (0,1)$. From (ii) it follows that $\lambda x_1 + (1 - \lambda)x_2 \in C$ for each $\lambda \in C$ for each $\lambda \in (0,1)$ and hence C is a convex cone.

Some examples of convex cones are:

(i) Hyperplane through a point x_o, i.e., $H = \{x : \langle x - x_o, p\rangle = 0\}$ where p is a nonzero vector in E_n. Here the vertex is x_o.

(ii) Closed half-spaces, i.e., $H^+ = \{x : \langle x - x_o, p\rangle \geq 0\}$ where p is a nonzero vector in E_n. Here again the vertex is x_o.

(iii) $C = \{x : Ax \geq 0\}$ where A is an mxn matrix.

The following lemma can easily be proved.*

3.1.5 Lemma. Let C_1 and C_2 be convex cones in E_n. Then $C_1 \cap C_2$, and $C_1 + C_2$ are also convex cones. Also if C is a cone in E_n then [C] is a convex cone in E_n.

The following gives the relationship between the sum of the convex cones and their union.

3.1.6 Lemma. Let C_1 and C_2 be convex cones in E_n. Then $C_1 + C_2 = [C_1 \cup C_2]$.

Proof: $x \in C_1 + C_2$ implies that $x = x_1 + x_2$ with $x_1 \in C_1$ and $x_2 \in C_2$. Hence $x_1, x_2 \in [C_1 \cup C_2]$ and since $[C_1 \cup C_2]$ is a convex cone then $x_1 + x_2 \in [C_1 \cup C_2]$. Hence $C_1 + C_2 \subset [C_1 \cup C_2]$. To prove the converse inclusion let $x \in [C_1 \cup C_2]$.

*In this chapter H(C), the convex hull of C, will also be denoted by [C].

Therefore $x = \sum_{i=1}^{k} \lambda_i x_i$ with $\lambda_i > 0$, $x_i \in C_1 \cup C_2$ $(i = 1,2,\ldots,k)$ and $\sum_{i=1}^{k} \lambda_i = 1$ where k is an integer greater or equal to one. Let $I_1 = \{i : x_i \in C_1\}$ and $I_2 = \{i \notin I_1 : x_i \in C_2\}$. Since C_1 and C_2 are convex then $y_1 = \sum_{i \in I_1} \lambda_i x_i \in C_1$ and $y_2 = \sum_{i \in I_2} \lambda_i x_i \in C_2$ and hence $x = y_1 + y_2 \in C_1 + C_2$. This completes the proof.

3.1.7 Definition. Let K be a nonempty set in E_n. Then the polar of K denoted by K^* is given by $K^* = \{p : \langle x,p \rangle \leq 0$ for each $x \in K\}$. If K is empty then K^* is interpreted as E_n.

In particular if we let K be a cone then we get the polar of a cone. Needless to say if the cone under consideration is a subspace L then $L^* = L^{\perp}$. Some examples of polar cones are given in Figure 3.1.

From the above definition of a polar cone, the following properties are obvious.

3.1.8 Lemma. Let K_1 and K_2 be nonempty sets (in particular cones) in E_n. Then

(i) K_1^* is a closed convex cone with vertex zero.

(ii) $K_1^* = (C\ell K_1)^*$

(iii) $K_1 \subset K_2$ implies $K_2^* \subset K_1^*$.

We will now discuss some properties of polar cones.

3.1.9 Theorem. Let K_1 and K_2 be nonempty cones in E_n. Then $(K_1 + K_2)^* = K_1^* \cap K_2^* = (K_1 \cup K_2)^*$.

Proof: We will first prove that $(K_1 + K_2)^* \supset K_1^* \cap K_2^*$. Let $p \in K_1^* \cap K_2^*$ then $\langle x,p \rangle \leq 0$ for each $x \in K_1$ and $\langle y,p \rangle \leq 0$ for each $y \in K_2$. Now let $z \in K_1 + K_2$, i.e., $z = x + y$ with $x \in K_1$ and $y \in K_2$. Hence $\langle z,p \rangle = \langle x,p \rangle + \langle y,p \rangle \leq 0$. Hence $p \in (K_1 +_2)^*$ and $(K_1 + K_2)^* \supset K_1^* \cap K_2^*$. To prove the converse inclusion let $p \in (K_1 + K_2)^*$. This means that $\langle z,p \rangle \leq 0$ for each $z \in K_1 + K_2$. Now consider $x \in K_1$ and $y \in K_2$. We claim that $\langle x,p \rangle \leq 0$ because otherwise $\langle \lambda x,p \rangle + \langle y,p \rangle > 0$ for λ sufficiently large which contradicts the fact that $\lambda x + y \in K_1 + K_2$ and $p \in (K_1 + K_2)^*$. Therefore $p \in K_1^*$ and similarly $p \in K_2^*$. Therefore $(K_1 + K_2)^* = K_1^* \cap K_2^*$.

To prove the second part of the theorem note that $K_1 \cup K_2 \supset K_1$ and so by Remark 3.1.8 we get $(K_1 \cup K_2)^* \subset K_1^*$. Similarly $(K_1 \cup K_2)^* \subset K_2^*$ and so

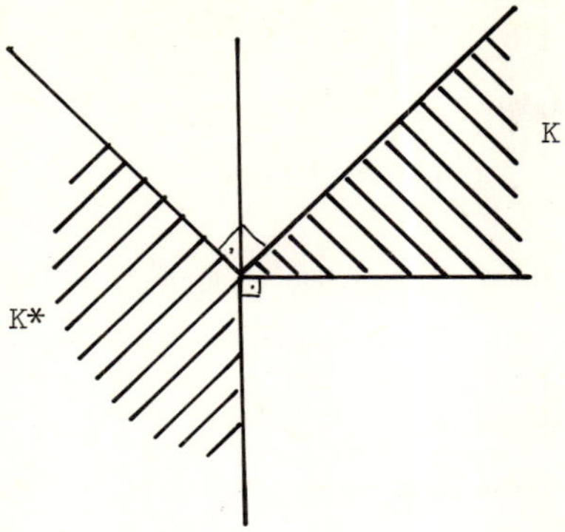

(i) $K = \{(x,y) : 0 \leq y \leq x\}$

$K^* = \{(x,y) : x \leq y \leq 0\}$

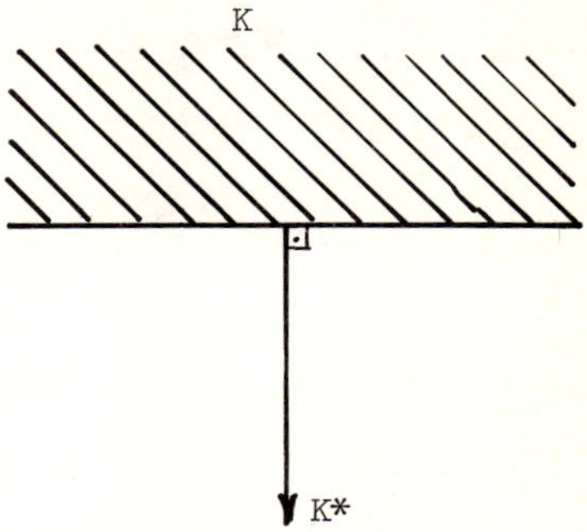

(ii) $K = \{(x,y) : y \geq 0\}$

$K^* = \{(x,y) : x = 0,\ y \leq 0\}$

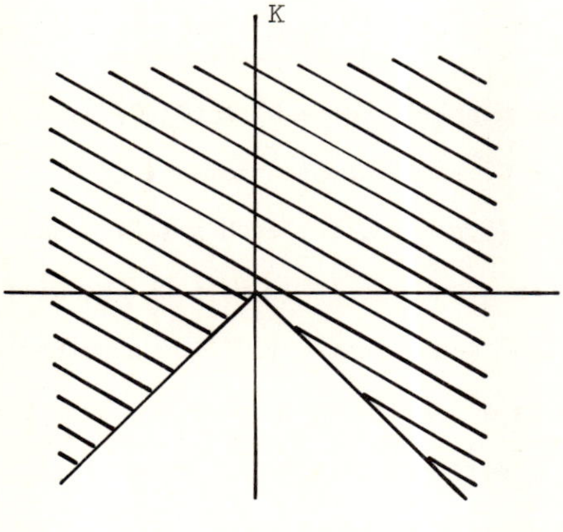

(iii) $K = \{(x,y) : y \geq -|x|\}$

$K^* = \{0\}$

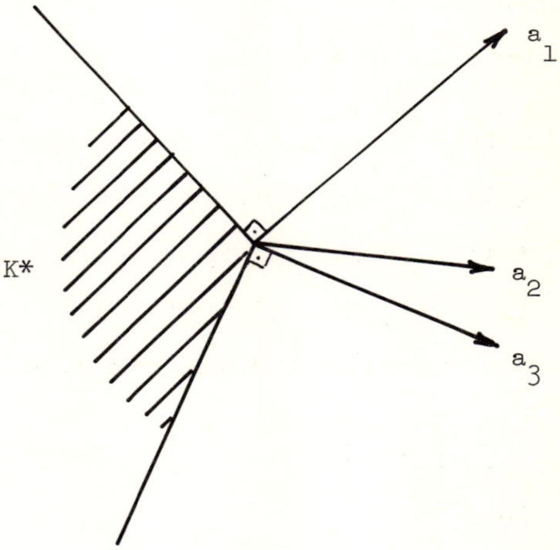

(iv) $K = \{x : x = Ap,\ p \geq 0\}$

$K^* = \{x : xA \leq 0\}$

where A is a mxn matrix, $p \in E^n$,

$x \in E^m$

Figure 3.1

Examples of Polar Cones

$(K_1 \cup K_2)^* \subset K_1^* \cap K_2^*$. On the other hand let $p \in K_1^* \cap K_2^*$, i.e., $\langle x,p \rangle \le 0$ for all $x \in K_1$ and $\langle x,p \rangle \le 0$ for all $x \in K_2$. Hence $\langle x,p \rangle \le 0$ for all $x \in K_1 \cup K_2$, i.e., $p \in (K_1 \cup K_2)^*$. This completes the proof.

Corollary: For linear subspaces L_1 and L_2, $(L_1 + L_2)^\perp = L_1^\perp \cap L_2^\perp$.

3.1.10 Theorem. Let K_1 and K_2 be arbitrary cones in E_n. Then $(K_1 \cap K_2)^* \supset K_1^* + K_2^* = [K_1^* \cup K_2^*]$.

Proof: If either K_1 or K_2 is empty then the result holds trivially. We now show that $(K_1 \cap K_2)^* \supset K_1^* + K_2^*$. Let $p \in K_1^* + K_2^*$, i.e., $p = p_1 + p_2$ with $p_1 \in K_1^*$ and $p_2 \in K_2^*$. Therefore $\langle x,p_1 \rangle \le 0$ for each $x \in K_1$ and $\langle x,p_2 \rangle \le 0$ for each $x \in K_2$. Now let $x \in K_1 \cap K_2$, then $\langle x,p \rangle = \langle x,p_1 \rangle + \langle x,p_2 \rangle \le 0$ and $p \in (K_1 \cap K_2)^*$. The second part of the Theorem follows from Lemma 3.1.6 by noting that K_1^* and K_2^* are convex cones.

We will now focus attention on convex cones.

3.1.11 Theorem. Let C be a nonempty convex cone in E_n. Then $C^{**} = C\ell\ C$.

Proof: Let $x \in C\ell\ C$. Then $\langle x,p \rangle \le 0$ for each $p \in C^*$ (note $C^* = (C\ell\ C)^*$) and hence $x \in C^{**}$. To prove that $C^{**} \subset C\ell\ C$ let $x \in C^{**}$ and suppose that $x \notin C\ell\ C$. By Lemma 2.4.6 there exists a nonzero p such that $\langle y,p \rangle \le \alpha$ for each $y \in C\ell\ C$ and $\langle x,p \rangle > \alpha$ for some α. But since $y = 0 \in C\ell\ C$ then $\alpha \ge 0$ and so $\langle x,p \rangle > 0$. We will show that this is impossible by showing that $p \in C^*$ (note that $x \in C^{**}$ by hypothesis). Suppose by contradiction that $p \notin C^*$ then there is a $y \in C$ with $\langle y,p \rangle > 0$. But then $\langle \lambda y,p \rangle$ can be made arbitrarily large by picking λ sufficiently large, which violates the fact that $\langle \lambda y,p \rangle \le \alpha$ for each $\lambda > 0$. This completes the proof.

In the special case when the cone under consideration is a linear subspace then we get $L = L^{\perp\perp}$.

3.1.12 Theorem. Let C_1 and C_2 be closed convex cones in E_n with nonempty intersection. Then $(C_1 \cap C_2)^* = C\ell\ (C_1^* + C_2^*) = C\ell\ [C_1^* \cup C_2^*]$.

Proof: From the above Theorem we have $C_1 = C_1^{**}$ and $C_2 = C_2^{**}$. Hence $(C_1 \cap C_2)^* = (C_1^{**} \cap C_2^{**})^*$. But from Theorem 3.1.9 it is true that $(C_1^* + C_2^*)^{**} = (C_1^{**} \cap C_2^{**})^*$ and the result then follows by Theorems 3.1.11 and Lemma 3.1.6.

Corollary 1. Let $\{C_i\}_{i \in I}$ be a finite collection of closed convex cones with nonempty intersection. Then $(\bigcap_{i\in I} C_i)^* = C\ell\,[\bigcup_{i\in I} C_i^*]$.

Proof: By finite induction of Theorem 3.1.12.

Corollary 2. Let $\{C_i\}_{i \in I}$ be a finite collection of closed convex cones. Then $\bigcap_{i\in I} C_i = \{0\}$ if and only if $[\bigcup_{i\in I} C_i^*] = E_n$.

Proof: Let $\bigcap_{i\in I} C_i = \{0\}$ then $E_n = \{0\}^* = (\bigcap_{i\in I} C_i)^* = C\ell\,[\bigcup_{i\in I} C_i^*]$ by Corollary 1. Hence $[\bigcup_{i\in I} C_i^*] = E_n$. Conversely if $E_n = [\bigcup_{i\in I} C_i^*] = C\ell[\bigcup_{i\in I} C_i^*]$ then $\{0\} = (E_n)^* = (C\ell\,[\bigcup_{i\in I} C_i^*])^* = (\bigcap_{i\in I} C_i)^{**} = \bigcap_{i\in I} C_i$. This completes the proof.

If we relax the assumption that C_1 and C_2 are closed, the Theorem does not hold in general even if C_1 and C_2 are convex cones with nonempty intersection. For example, let $C_1 = \{(x,y) : x > 0,\ y > 0\} \cup \{0,0\}$ and $C_2 = \{(x,y) : x < 0,\ y > 0\} \cup \{0,0\}$. Then $C_1 \cap C_2 = \{0,0\}$ and $(C_1 \cap C_2)^* = E_2$. On the other hand $C_1^* = \{(x,y) : x \le 0,\ y \le 0\}$ and $C_2^* = \{(x,y) : x \ge 0,\ y \le 0\}$ and hence $C\ell\,(C_1^* + C_2^*) = C_1^* + C_2^* = \{(x,y) : x \in E_1,\ y \le 0\}$ which is obviously not equal to E_2.

Furthermore under the assumptions of the above Theorem we cannot claim in general that $(C_1 \cap C_2)^* = C_1^* + C_2^*$. The problem is that $C_1^* + C_2^*$ is not closed in general whereas $(C_1 \cap C_2)^*$ is alwasy closed. The following is a sufficient condition for $C_1^* + C_2^*$ to be closed. It turns out that the same condition is sufficient for $C\ell\,(C_1 \cap C_2) = C\ell\, C_1 \cap C\ell\, C_2$ to hold.

3.1.13 Lemma. Let C_1 and C_2 be convex cones in E_n with $riC_1 \cap riC_2 \ne \emptyset$. Then $C\ell\,(C_1 \cap C_2) = C\ell\, C_1 \cap C\ell\, C_2$ and $C_1^* + C_2^*$ is closed.

Proof: It is obvious that $C\ell\,(C_1 \cap C_2) \subset C\ell\, C_1 \cap C\ell\, C_2$. To show the converse let $x \in C\ell\, C_1 \cap C\ell\, C_2$ and $x_0 \in riC_1 \cap riC_2$ (such an x_0 exists by hypothesis). By Theorem 2.3.5 it follows that $x_\lambda = \lambda x_0 + (1 - \lambda)x \in riC_1 \cap riC_2 \subset C_1 \cap C_2$ for each $\lambda \in (0,1)$. Therefore $x = \lim_{\lambda\to 0} x_\lambda \in C\ell\,(C_1 \cap C_2)$.

To prove the second part let $u \in C\ell\,(C_1^* + C_2^*)$. To show that $C_1^* + C_2^*$ is closed we then need to show that u is the sum of two vectors in C_1^* and C_2^* respectively. $u \in C\ell\,(C_1^* + C_2^*)$ means that there is a sequence $a_k + b_k \to u$ with $a_k \in C_1^*$ and $b_k \in C_2^*$. Let L_1 and L_2 be the linear subspaces generated by C_1 and C_2 repsectively, and let $L_1^\perp$ and $L_2^\perp$ be their orthogonal complements. By the projection Theorem 1.3.3 a_k and b_k can be uniquely represented as $a_k = a_k' + a_k^\perp$ and $b_k = b_k' + b_k^\perp$ where $a_k' \in L_1$, $b_k' \in L_2$,

$a_k^{\perp} \in L_1^{\perp}$, and $b_k^{\perp} \in L_2^{\perp}$. Let $x_o \in riC_1 \cap riC_2$. The sequence $\langle x_o, a_k\rangle$ is bounded above by zero and is also bounded below since $\langle x_o, b_k\rangle$ is bounded above and $\langle x_o, a_k + b_k\rangle \to \langle x_o, u\rangle$. Hence $\{\langle x_o, a_k\rangle\}$ has a convergent subsequence $\langle x_o, a_{k_j}\rangle \to \alpha$. Since $x_o \in riC_1$ then there is an $\epsilon > 0$ with $L_1 \cap N_\epsilon(x_o) \subset C_1$ (note that for a cone, the affine manifold and its parallel subspace and the subspace spanned by the cone are equal). This means that for every nonzero a'_{k_j} we must have $x_o + \epsilon/2 \dfrac{a'_{k_j}}{\|a'_{k_j}\|} \in C_1$ and and hence we get $0 \geq \langle x_o + \epsilon/2 \dfrac{a'_{k_j}}{\|a'_{k_j}\|}, a_{k_j}\rangle = \langle x_o + \epsilon/2 \dfrac{a'_{k_j}}{\|a'_{k_j}\|}, a'_{k_j}\rangle = \langle x_o, a'_{k_j}\rangle + \epsilon/2 \|a'_{k_j}\|$. This implies that $\epsilon/2 \|a'_{k_j}\| \leq -\langle x_o, a'_{k_j}\rangle$ and since the latter converges to $-\alpha$ (note that $\langle x_o, a'_{k_j}\rangle = \langle x_o, a_{k_j}\rangle$) then $\|a'_{k_j}\|$ is bounded and has a convergent subsequence. In other words, $a'_{k_{j_i}} \to a'$. Since $a'_{k_{j_i}} \in L_1$ and L_1 is closed by Theorem 1.2.6, then $a' \in L_1$. Therefore $b'_{k_{j_i}} + a^{\perp}_{k_{j_i}} + b^{\perp}_{k_{j_i}} \to u - a'$. In a similar fashion we can show that $b'_{k_{j_i}}$ has a convergent subsequence $b'_{k_{j_{i_\ell}}} \to b' \in L_2$ and hence we get $a^{\perp}_{k_{j_{i_\ell}}} + b^{\perp}_{k_{j_{i_\ell}}} \to u - a' - b'$. Since $a^{\perp}_{k_{j_{i_\ell}}} + b^{\perp}_{k_{j_{i_\ell}}} \in L_1^{\perp} + L_2^{\perp}$ which is a subspace and hence closed then $u - a' - b' \in C\ell\,(L_1^{\perp} + L_2^{\perp}) = L_1^{\perp} + L_2^{\perp}$. Now let $u - a' - b' = c + d$ with $c \in L_1^{\perp}$ and $d \in L_2^{\perp}$. Note that $a' + c \in C_1^*$ because a' is the limit of $a'_{k_{j_{i_\ell}}}$ where $\langle x, a'_{k_{j_{i_\ell}}}\rangle \leq 0$ for each $x \in C_1$ and $c \in L_1^{\perp}$. Similarly $b' + d \in C_2^*$. Hence $u = (a' + c) + (b' + d) \in C_1^* + C_2^*$. This completes the proof.

We will now make use of the closedness of $C_1^* + C_2^*$ to show that $(C_1 \cap C_2)^* = C_1^* + C_2^*$. This result is a very important result and will be used later.

3.1.14 Theorem. Let C_1 and C_2 be convex cones in E_n with $riC_1 \cap riC_2 \neq \emptyset$. Then $(C_1 \cap C_2)^* = C_1^* + C_2^*$.

Proof: By Lemma 3.1.13 we get $C_1^* + C_2^* = C\ell\,(C_1^* + C_2^*)$. But by Theorem 3.1.11 it follows that $C\ell\,(C_1^* + C_2^*) = (C_1^* + C_2^*)^{**}$ and hence $C_1^* + C_2^* = (C_1^* + C_2^*)^{**} = (C_1^{**} \cap C_2^{**})^*$ by Theorem 3.1.9. But applying Theorem 3.1.11 and Lemma 3.1.13 we get $C_1^* + C_2^* = (C_1^{**} \cap C_2^{**})^* = (C\ell\, C_1 \cap C\ell\, C_2)^* = (C\ell\,(C_1 \cap C_2))^*$. The result then follows by noting Lemma 3.1.8 part (ii).

For the special case of linear subspaces we get $(L_1 \cap L_2)^{\perp} = L_1^{\perp} + L_2^{\perp}$. Note that the condition $\text{ri}L_1 \cap \text{ri}L_2 \neq \emptyset$ is satisfied since $0 \in \text{ri}L_1 \cap \text{ri}L_2$.

3.2. Polyhedral Cones

We will now discuss a special class of cones, namely polyhedral cones. These cones are generated by a finite number of points and always contain the origin.

3.2.1 Definition. C is said to be a polyhedral cone if there exists a matrix A such that $C = \{Ax : x \geq 0\}$.

Note that 0 is an element of every polyhedral cone. Note also that a polyhedral cone is convex. We will show later that a polyhedral cone is always closed. Some examples of polyhedral cones are:

(i) $C = \{0\}$

(ii) $C = \{\lambda(1,0) : \lambda \geq 0\}$

(iii) $C = \{\lambda_1(1,0) + \lambda_2(2,2) : \lambda_1, \lambda_2 \geq 0\}$

(iv) $C = \{x : \langle p, x \rangle = 0\}$ where p is a nonzero vector in E_n.

We may interpret C as the sum of the rays formed by the columns of A. Alternatively, C is the nonnegative combinations of the columns of A. Hence corresponding to the points $x_1, x_2, \ldots, x_k$ we can construct a polyhedral cone C as $C(x_1, x_2, \ldots, x_n) \cup \{0\}$.

3.2.2 Theorem. Let $K = \{Ax : x \geq 0\}$ be a polyhedral cone and let $C = \{y : A^t y \leq 0\}$. Then $C = K^*$.

Proof: Consider $y \in C$ then for $z \in K, y^t z = y^t Ax \leq 0$ since $y^t A \leq 0$ and $x \geq 0$. Hence $C \subset K^*$. To show that $K^* \subset C$ let $y \in K^*$. Then $y^t Ax \leq 0$ for each $x \geq 0$. This implies that $y^t A \leq 0$ since otherwise some $y^t a_j > 0$, where a_j is the jth column of A, and $y^t Ax$ can be made positive by choosing the jth component of $x > 0$ while all other components zero. Hence $y^t A \leq 0$, i.e., $y \in C$. This shows that $K^* \subset C$ and the proof is complete.

The above theorem gives an explicit expression for the polar of a polyhedral cone. The theorem also shows that $C^* = K^{**}$. Later we will show that K is always closed and hence $C^* = K$. We would like to show that the polar of a polyhedral cone is indeed a polyhedral cone. In order to do that we essentially need to show that the system $Ax \leq 0$ is a polyhedral cone according to Definition 3.2.1. This will be done by making use of the following lemma.

3.2.3 Lemma. The set $C = \{x : x \geq 0, A^t x = 0\}$ is a polyhedral cone.

Proof: Consider $S = \left\{x : x \geq 0, A^t x = 0, \sum_{i=1}^{n} x^i = 1\right\}$ where x^i is the ith component of the vector x. Note that S can be represented as $\{x : Dx \geq b\}$ where $D^t = (I,A,-A,e,-e)$ and $b^t = (0,0,0,1,1)$ and e is the n vector with each component equal to 1. Note that S is not empty as long as $C \neq \{0\}$. As a matter of fact a nonzero $x \in C$ if and only if $x/\sum_{i=1}^{n} x^i \in S$. If $C = \{0\}$ then it is indeed a polyhedral cone. If $C \neq \{0\}$ then $S \neq \emptyset$ and since S is bounded then by Theorem 2.4.12 it is obvious that S has a finite number of extreme points $b_1, b_2, \ldots, b_k$ and their convex hull is the set S itself. Since any $x \in C$ can be represented as μy where $y \in S$ and $\mu \geq 0$ and since y can be represented as a convex combination of $b_1, b_2, \ldots, b_k$ it is then immediate that $x = \sum_{i=1}^{k} \lambda_i b_i$ where $\lambda_i \geq 0$. In other words $C \subset \{x : x = B\lambda, \lambda \geq 0\}$ where $B = (b_1, b_2, \ldots, b_k)$. Conversely let x be such that $x = B\lambda$ where $\lambda \geq 0$. Suppose that $\lambda \neq 0$. Then $x/\sum_{i=1}^{n} \lambda_i$ is a convex combination of $b_1, b_2, \ldots, b_k$ and hence belongs to S. Therefore x belongs to C and we get $C = \{B\lambda : \lambda \geq 0\}$. This shows that C is indeed a polyhedral cone.

Corollary: Let L be a given linear subspace. Then $C = \{x \in L : x \geq 0\}$ is a polyhedral cone.

Proof: Consider the orthogonal complement $L^\perp$ of L. By Theorem 1.2.4, $L^\perp = \{By : y \in E_m\}$ where $B = (b_1, b_2, \ldots, b_m)$ is an nxm matrix corresponding to the basis of $L^\perp$. Let $D = \{x : B^t x = 0, x \geq 0\}$. We claim that $D = C$. First let $x \in C$. Since $b_i \in L^\perp$ and $x \in L$ then $\langle x, b_i \rangle = 0$ for each i and hence $B^t x = 0$, i.e., $x \in D$. Now let $x \in D$ and let $y \in L^\perp$. Since $y \in L^\perp$ and $B = (b_1, b_2, \ldots, b_m)$ is a basis of $L^\perp$ then $y = Bz$ for some vector z. But this shows that $\langle x, y \rangle = \langle x, Bz \rangle = 0$ since $x \in D$ means that $B^t x = 0$. Therefore $x \in L^{\perp\perp} = L$ by Theorem 3.1.11. This shows that $x \in C$, i.e., $C = D$. This completes the proof.

3.2.4 Theorem. Let $C = \{x : Ax \leq 0\}$ where A is an mxn matrix. Then C is a polyhedral cone.

Proof: We need to show that C can be written as $\{Fz : z \geq 0\}$ where F is an nxk matrix. Now consider the following set: $S = \{y : y \geq 0, y = Ax, x \in E_n\}$. S is the intersection of the nonnegative orthant and the linear subspace given by $\{y : y = Ax$ for some $x \in E_n\}$. By the Corollary to Lemma 3.2.3 above it is clear that S is a polyhedral cone and hence $S = \{Bz : z \geq 0\}$ where B is an mxk_1 matrix with columns $b_1, b_2, \ldots, b_{k_1}$. Since each $b_i \in S$ (e.g., by picking z to be the vector of zeros and 1 in the ith position) then $b_i \geq 0$ and $b_i = Ax_i$ for some x_i. In other words $B = AD$ where D is an nxk_1 matrix with columns $x_1, x_2, \ldots, x_{k_1}$. Now consider the linear subspace $L = \{x : Ax = 0\}$. By Theorem 1.2.4, L can be represented as $L = \{Ez : z \geq 0\}$ where E is nxk_2 matrix. We claim that C is the polyhedral cone represented by $\{Fz : z \geq 0\}$ where $F = (-D, E)$ as an nxk matrix with $k = k_1 + k_2$. To show this let $x \in C$, i.e., $0 \leq y = -Ax$. This means that $y \in S$ and hence $y = Bz_1$ for some k_1 vector $z_1 \geq 0$. But then $-Ax = y = ADz_1$, i.e., $A(x + Dz_1) = 0$. This means that $x + Dz_1 \in L$ and hence $x + Dz_1 = Ez_2$ for some k_2 vector $z_2 \geq 0$. This means that $x = -Dz_1 + Ez_2 = Fz$ with $z^t = (z_1^t, z_2^t) \geq 0$, i.e., $C \subset \{Fz : z \geq 0\}$. Conversely consider the vector $Fz = -Dz_1 + Ez_2$ with $z_1, z_2 \geq 0$. Then $AFz = -ADz_1 + AEz_2$. But note that $Ez_2 \in L$ as long as $z_2 \geq 0$ and so $AEz_2 = 0$. Also note that $AD = B$ and so $-ADz_1 = -Bz_1$. But $Bz_1 \in S$ and so $Bz_1 \geq 0$ and $-Bz_1 \leq 0$. This shows that $AFz \leq 0$ and so $\{Fz : z \geq 0\} \subset C$ and the proof is complete.

The above theorem together with Theorem 3.2.2 essentially say that the polar of a polyhedral cone is another polyhedral cone. This actually gives us the flexibility of interchanging the definition of a polyhedral cone between $C = \{Ax : x \geq 0\}$ or as $C = \{x : Bx \leq 0\}$. As a matter of fact the above theorem coupled with Theorem 3.2.2 shows that any set of the form $\{Ax : x \geq 0\}$ can be represented in a dual form as $\{x : Bx \leq 0\}$.

To illustrate this geometrically appealing fact let us consider the nonnegative linear combination of the two vectors $a_1 = (1, 2)$ and $a_2 = (-1, 1)$, i.e., the set of points $(x_1 - x_2, 2x_1 + 2x_2)$ where x_1 and x_2 are nonnegative real numbers. We can represent the set $\{(x_1 - x_2, 2x_1 + 2x_2) : x_1, x_2 \geq 0\}$ as the set $\{x : Bx\}$ where B is an appropriate matrix which can be found, theoretically at least, using the constructive proof of Theorem 3.2.4 above. Let us consider the vectors (2, -1) and (-1, -1). Note that $(2, -1) \perp (1, 2)$ and $(-1, -1) \perp (-1, 1)$. Consider the vectors

(y_1, y_2) such that $Bx \le 0$. We must then have $2y_1 - y_2 \le 0$ and $-y_1 - y_2 \le 0$. We leave it for the reader to check that the sets $\{(y_1, y_2) : 2y_1 - y_2 \le 0, -y_1 - y_2 \le 0\}$ and $\{(x_1 - x_2, 2x_1 + 2x_2) : x_1 \ge 0, x_2 \ge 0\}$ are equal. These two dual representations of the same set are illustrated in Figure 3.2 below.

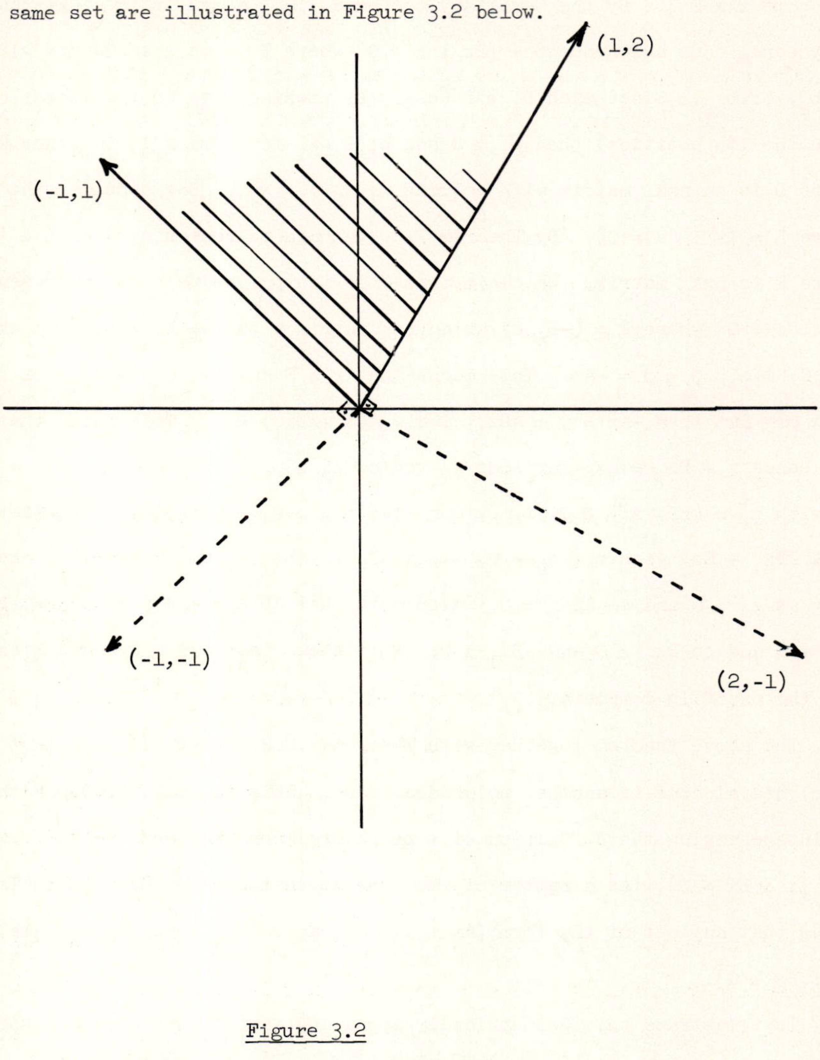

Figure 3.2

Dual Representation of a Polyhedral Cone

We have noted earlier that a polyhedral cone C is convex. From the discussion following Theorem 3.2.4 we have also seen that any polyhedral cone can be represented as $C = \{x : Bx \le 0\}$. That is, C can be represented as the intersection of closed half spaces. Hence C is also closed. Finally, 0 is an element of every polyhedral cone. To summarize, a polyhedral cone is always a nonempty closed and convex cone. The following important theorem then readily follows from Theorem 3.1.11.

3.2.5 Theorem. If C is a polyhedral cone then $C = C^{**}$.

The following theorem summarizes some of the important facts about polyhedral cones. Some of these results hold true for closed convex cones in general but some others do not.

3.2.6 Theorem. Let C_1 and C_2 be polyhedral cones. Then

(i) $C_1 + C_2$ is a polyhedral cone.

(ii) C_1^* and C_2^* are polyhedral cones.

(iii) $C_1 \cap C_2$ is a polyhedral cone.

(iv) $(C_1 + C_2)^* = C_1^* \cap C_2^*$.

(v) $(C_1 \cap C_2)^* = C_1^* + C_2^*$.

Proof: (i) is immediate from definition and (iv) is true for all cones (see Theorem 3.1.9). (ii) follows from Theorems 3.2.2 and 3.2.4. Now we prove (iii). By Theorem 3.2.5 above $C_1 \cap C_2 = C_1^{**} \cap C_2^{**} = (C_1^* + C_2^*)^*$ from Theorem 3.1.9. But by (ii) C_1^* and C_2^* are polyhedral and so is $C_1^* + C_2^*$. Again by (ii) $(C_1^* + C_2^*)^*$ is polyhedreal and so is $C_1 \cap C_2$. Finally we show (v). $C_1^* + C_2^* = (C_1^* + C_2^*)^{**} = (C_1^{**} \cap C_2^{**})^* = (C_1 \cap C_2)^*$. This completes the proof.

As a special case part (v) of the above theorem we can develop Farkas Lemma. Let $C_i = \{x : \langle x, a_i\rangle \leq 0\}$ for $i = 1,2,\ldots,m$. In other words each C_i represents a closed halfspace with normal vector a_i. Note that $C = \bigcap_{i=1}^{m} C_i = \{x : A^t x \leq 0\}$ where A is the $n \times m$ matrix with columns $a_1, a_2, \ldots,$ and a_m. This then means that C^* composes of vectors b such that $\langle b, x\rangle \leq 0$ whenever $A^t x \leq 0$, i.e., $C^* = \{b : A^t x \leq 0$ implies $\langle x, b\rangle \leq 0\}$. It is clear by definition of each C_i that $C_i^* = \{\lambda a_i : \lambda \geq 0\}$. Therefore applying part (v) of Theorem 3.2.6 we get the following result: the system $A^t x \leq 0$ implies that $\langle x, b\rangle \leq 0$ is consistent if and only if the system $b = A\lambda$ has a solution with $\lambda \geq 0$. This is precisely Farkas Lemma.

It may be noted that Farkas Lemma can also be stated as follows: The following two systems are equivalent:

I. $A^t x \leq 0$ implies $\langle b, x \rangle \leq 0$,

II. $b = A\lambda$ with $\lambda \geq 0$.

Farkas Lemma can be put in an equivalent form as follows: Either system I or II below has a solution but never both.

I. $A^t x \leq 0$ with $\langle b, x \rangle > 0$,

II. $b = A\lambda$ with $\lambda \geq 0$.

3.3. Cones Generated by Sets

In Chapter 1 we have already introduced the notion of a convex cone generated by an arbitrary set. The cone defined therein was not only convex but had its vertex at zero. In this section we will introduce different cones associated with a set which may not be convex. The vertex of each of these cones is again the zero vector by a suitable translation operation. We will also discuss several properties of these cones.

3.3.1 Definition. Let S be a nonempty set in E_n and let $x_o \in E_n$. Then $K(S,x_o)$ is said to be cone of S at x_o if it contains $S - x_o$ and is contained in all such cones.

This cone will sometimes be referred to as the cone generated by $S - x_o$. So given a set S and a point x_o, in order to find $K(S, x_o)$, we first construct the set $S - x_o$ and then find the cone generated by $S - x_o$. Therefore it is clear that changing the point x_o will give rise to completely different cones. Figure 3.3 shows some examples of cones generated by a set S at a given point x_o. Clearly such cones need not be convex.

From the above definition it is clear that if x_o is an interior point of S then $K(S, x_o)$ is indeed E_n. This follows from the fact that $S - x_o$ will have 0 as an interior point and any cone with vertiex zero containing $S - x_o$ is E_n. The following remark gives an important characterization of cones generated by an arbitrary set via rays emanating from the point x_o through the set S under consideration. By examining Figure 3.3 the reader may note that any ray in $K(S, x_o)$ must meet the set $S - x_o$ at some point. This is proved below.

3.3.2 Remark. Let S be nonempty set in E_n and $x_o \in E_n$. Then $K(S, x_o) = \{\lambda(x - x_o) : x \in S, \lambda > 0\}$.

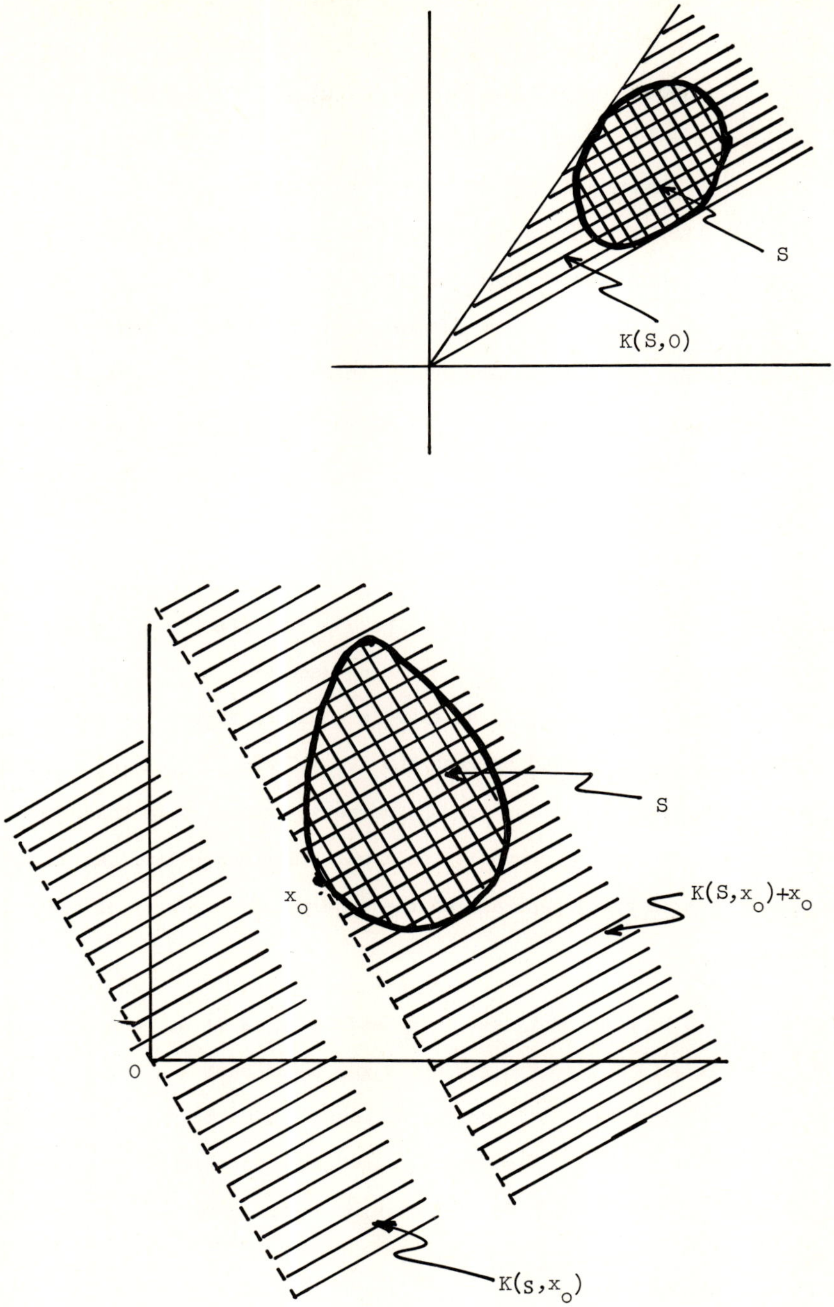

Figure 3.3

Examples of Cones Generated by Sets

<u>Proof</u>: The set $\{\lambda(x - x_o) : x \in S, \lambda > 0\}$ is obviously a cone with vertex zero. It also contains $S - x_o$ by letting $\lambda = 1$. To show that $K(S, x_o)$ is the smallest such cone, let K be a cone that contains $S - x_o$. Now $y \in K(S, x_o)$ implies that $y = \lambda(x - x_o)$ for some $x \in S$ and $\lambda > 0$. But since $S - x_o \subset K$ then $x - x_o \in K$ and so is $y = \lambda(x - x_o)$. Hence $K(S, x_o) \subset K$ and the proof is complete.

We will interpret $K(S, x_o)$ when S is empty as the empty set, regardless of x_o. It may be noted from the above remark that $0 \in K(S, x_o)$ if and only if $x_o \in S$. In some cases one would like to find the cone generated by the intersection of two sets (and the union of two sets) in terms of the cones generated by the individual sets. The following Lemma asserts that the cone of the union is the union of the cones and that the cone of the intersection is a subset of the intersection of the cones. The proof of the lemma is trivial.

3.3.3 <u>Lemma</u>. Let S_1 and S_2 be arbitrary sets in E_n and let $x_o \in E_n$. Then $K(S_1 \cup S_2, x_o) = K(S_1, x_o) \cup K(S_2, x_o)$ and $K(S_1 \cap S_2, x_o) \subset K(S_1, x_o) \cap K(S_2, x_o)$.

The converse inclusion corresponding to the cone generated by the intersection of two sets does not hold in general. As an example consider Figure 3.4 where $S_1 \cap S_2 = \{0\}$ and $K(S_1 \cap S_2, x_o) = \{0\}$ whereas $K(S_1, x_o) \cap K(S_2, x_o) \neq \{0\}$. Lemma 3.3.5 below gives a sufficient condition for the converse inclusion to hold. For this purpose we need the concept of convexity at a point defined below. Actually the notion of convexity at a point generalizes the concept of convexity of a set discussed in Chapter 2.

3.3.4 <u>Definition</u>. Let S be a nonempty set in E_n and let $x_o \in C\ell\, S$. Then S is said to be <u>convex at x_o</u> if $\lambda x + (1 - \lambda)x_o \in S$ for every $x \in S$ and each $\lambda \in (0,1)$.

Note that if S is convex then it is convex at each $x_o \in C\ell\, S$, and conversely. This follows from Definition 2.1.3. Figure 3.5 below gives an example of a nonconvex set which is convex at some point x_o. We will develop in the next section an important characteristic of convexity at a point.

3.3.5 <u>Lemma</u>. Let S_1 and S_2 be arbitrary sets in E_n which are convex at $x_o \in C\ell\, S_1 \cap C\ell\, S_2$. Then $K(S_1 \cap S_2, x_o) = K(S_1, x_o) \cap K(S_2, x_o)$.

<u>Proof</u>: Let $y \in K(S_1, x_o) \cap K(S_2, x_o)$, i.e., $y = \lambda_1(x_1 - x_o) = \lambda_2(x_2 - x_o)$

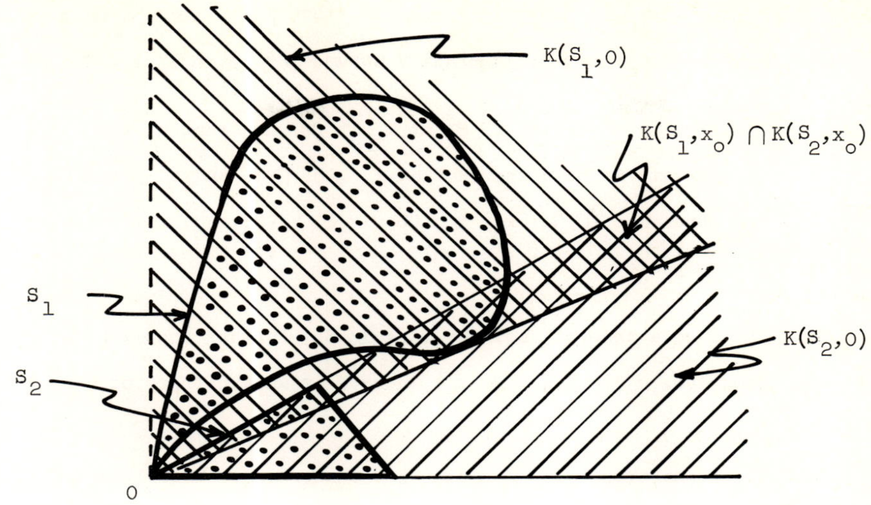

Figure 3.4

Cone Generated by the Intersection of Two Sets

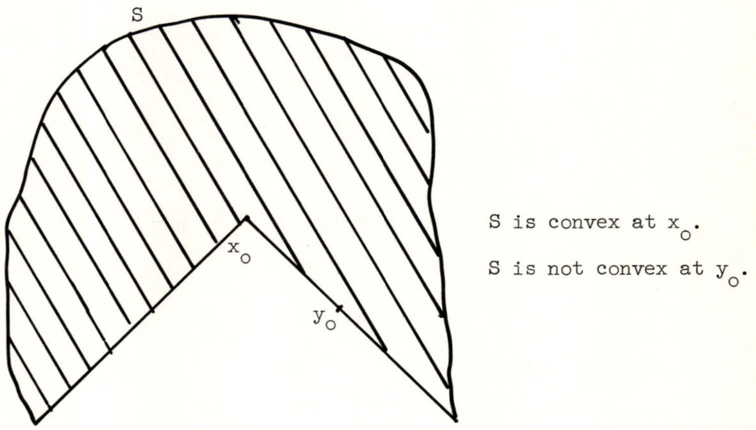

Figure 3.5

Convexity at a Point

where $\lambda_1, \lambda_2 > 0$ and $x_1 \in S_1$, $x_2 \in S_2$. If $\lambda_1 = \lambda_2$ then $x_1 = x_2$ and then $x_2 \in S_1 \cap S_2$. Otherwise without loss of generality assume that $\lambda_1 < \lambda_2$. Since $x_2 = \frac{\lambda_1}{\lambda_2} x_1 + \left(1 - \frac{\lambda_1}{\lambda_2}\right)x_o$ then by convexity of S_1 at x_o it follows that $x_2 \in S_1$, i.e., $x_2 \in S_1 \cap S_2$. At any rate we have showed that $x_2 \in S_1 \cap S_2$ and so $y = \lambda_2(x_2 - x_o) \in K(S_1 \cap S_2, x_o)$ and hence $K(S_1 \cap S_2, x_o) \supset K(S_1, x_o) \cap K(S_2, x_o)$. In view of Lemma 3.3.3 the proof is complete.

3.3.6 Lemma. Let S be a nonempty set and $x_o \in C\ell\, S$. Further suppose that S is convex at x_o. Then $K(S, x_o) = K(S \cap N, x_o)$ where N is an arbitrary open ball about x_o.

Proof: Let N be an open ball of radius $\epsilon > 0$ about x_o. Clearly $K(S \cap N, x_o) \subset K(S, x_o)$. Conversely, let $y \in K(S, x_o)$, then $y = \lambda(x - x_o)$ with $x \in S$, $\lambda > 0$. Without loss of generality assume that $\|x - x_o\| = k \geq \epsilon$. It is obvious that y can be represented as $y = \mu(z - x_o)$ where $\mu = \frac{2\lambda k}{\epsilon}$ and $z = \frac{\lambda}{\mu} x + \left(1 - \frac{\lambda}{\mu}\right) x_o$. By convexity of S at x_o and since $\frac{\lambda}{\mu} \in (0,1)$ it follows that $z \in S$. Moreover $\|z - x_o\| = \frac{\lambda}{\mu}\|x - x_o\| = \epsilon/2 < \epsilon$ and hence $z \in N$. Therefore $y \in K(S \cap N, x_o)$ and we have the desired result.

We will now introduce the concept of a convex cone of a set S at x_o denoted by $C(S, x_o)$ which will play an important role later. It will be readily recognized that $C(S)$, the convex cone of S (convex cone generated by A) defined in Chapter 2 is the convex cone of S at a point $x_o = 0$, i.e., when we suppress the point x_o it is assumed that $x_o = 0$ and the convex cone of S at $x_o = 0$ is simply called the convex cone of S (convex cone generated by S). We will follow this convention throughout, i.e., whenever convenient $C(S,0)$ is denoted by $C(S)$.

3.3.7 Definition. Let S be a nonempty set in E_n and $x_o \in E_n$. Then $C(S, x_o)$ is said to be the convex cone of S at x_o if it is a convex cone that contains $S - x_o$ and which is contained in all such cones.

If S is empty then we will interpret $C(S, x_o)$ as the empty set for each x_o. Figure 3.6 shows an example of the convex cone generated by a set S at the origin. The reader may note that this cone is the convex hull of the cone generated by S at the origin.

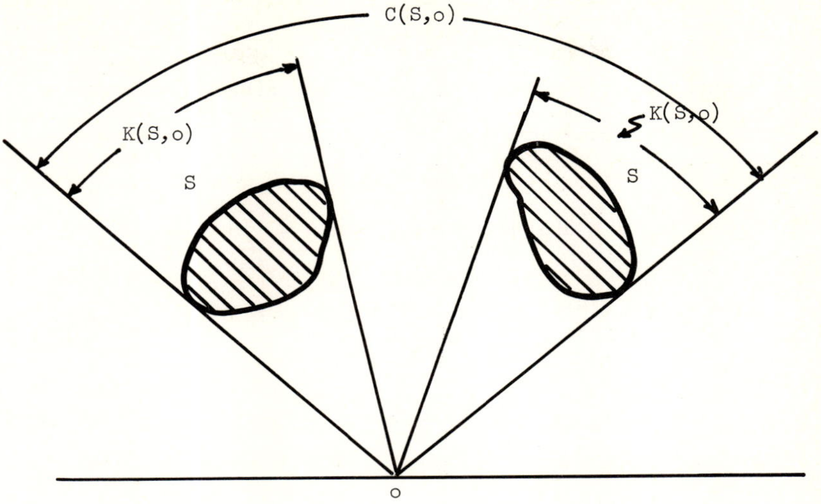

Figure 3.6

An Example of the Convex Cone of a Set

It will be seen from the following lemma that if S is convex then $K(S, x_o) = C(S, x_o)$, i.e., the cone of S at x_o is also the convex cone of S at x_o.

3.3.8 Lemma. Let S be a nonempty convex set in E_n. Then $K(S, x_o) = C(S, x_o)$.

Proof: If we show that $K(S, x_o)$ is convex then the result is at hand by definitions of $K(S, x_o)$ and $C(S, x_o)$. Let $y_1, y_2 \in K(S, x_o)$. It suffices to show that $y_1 + y_2 \in K(S, x_o)$. But $y_1 = \lambda_1(x_1 - x_o)$ and $y_2(x_2 - x_o)$ where $\lambda_1, \lambda_2 > 0$, $x_1, x_2 \in S$. Then $y_1 + y_2 = \lambda_1 x_1 + \lambda_2 x_2 - (\lambda_1 + \lambda_2)x_o = (\lambda_1 + \lambda_2)\left(\frac{\lambda_1}{\lambda_1 + \lambda_2} x_1 + \frac{\lambda_2}{\lambda_1 + \lambda_2} x_2 - x_o\right)$. But $\frac{\lambda_1}{\lambda_1 + \lambda_2} x_1 + \frac{\lambda_2}{\lambda_1 + \lambda_2} x_2 \in S$ by convexity of S. Hence $y_1 + y_2 \in K(S, x_o)$ and the proof is complete.

From the above result it is then chear that the cone of a convex set is indeed convex. If S is not convex then we can construct $C(S, x_o)$ by means of the following remark which is illustrated in Figure 3.6 above.

3.3.9 Theorem. Let S be a nonempty set in E_n and $x_o \in E_n$. Then $C(S, x_o) = K(H(S), x_o) = H(K(S, x_o))$.

Proof: We will prove the result by showing that $C(S, x_o) \subset K(H(S), x_o) \subset H(K(S, x_o)) \subset C(S, x_o)$. By Lemma 3.3.2 and 3.3.8, $K(H(S), x_o) = \{\lambda(x - x_o) : x \in H(S), \lambda > 0\}$ is a convex cone that contains $H(S) - x_o$ and hence $S - x_o$. Hence from Definition 3.3.7, $C(S, x_o) \subset K(H(S), x_o)$. If we now let $y \in K(H(S), x_o)$ then $y = \lambda(x - x_o)$ for some $x \in H(S)$ and $\lambda > 0$. Then $x = \sum_{i=1}^{k} \lambda_i x_i$ where $\lambda_i \geq 0$, $x_i \in S$, $\sum_{i=1}^{k} \lambda_i = 1$ and k is a positive integer. Therefore $y = \lambda(x - x_o) = \lambda\left(\sum_{i=1}^{k} \lambda_i x_i - x_o\right)$

$= \sum_{i=1}^{k} \lambda_i(\lambda(x - x_o))$ and hence $y \in H(K(S, x_o))$. Finally we show that $H(K(S, x_o)) \subset$

$C(S, x_o)$. Let $y \in H(K(S, x_o))$ then $y = \sum_{i=1}^{k} \lambda_i y_i$ where $\lambda_i \geq 0$, $y_i \in K(S, x_o)$, $\sum_{i=1}^{k} \lambda_i = 1$ and k is a positive integer. But since $K(S, x_o) \subset C(S, x_o)$ then $y_i \in C(S, x_o)$ and by convexity of the latter it follows that $y \in C(S, x_o)$. This completes the proof.

We will make use of the above theorem in finding the minimum convex cone with vertex (1,1,1) that contains the points (1,1,1), (2,1,2), and (3,4,6). To do this let $S = \{(1,1,1), (2,1,2), (3,4,6)\}$ and $x_o = (1,1,1)$. We first find the convex cone generated by the vectors (2,1,2) - (1,1,1) = (1,0,1) and (3,4,6) - (1,1,1) = (2,3,5) and then translate the vertex from the origin to the point (1,1,1). So the convex cone generated by (1,0,1) and (2,3,5) is $\{\lambda + 2\mu, 3\mu, \lambda + 5\mu : \lambda \geq 0, \mu \geq 0\}$. This means that the cone we are looking for is the set $\{1 + \lambda + 2\mu, 1 + 3\mu, 1 + \lambda + 5\mu : \lambda \geq 0, \mu \geq 0\}$. One can easily verify if any point belongs to the cone. For example let us consider the point (2,4,3). If this point belongs to the cone then $1 + 3\mu = 4$ and so $\mu = 1$. But if $\mu = 1$ then $\lambda = -1$ and so $(2,4,3) \notin C(S, x_o)$ since λ has to be nonnegative. On the other hand the point (7,7,13) belongs to $C(S, x_o)$ with $\lambda = \mu = 2$.

We will have occasion to use polars of the above mentioned cones. The following Theorem shows the equivalence of polars of sets and their cones and convex cones.

3.3.10 Theorem. Let S be a nonempty set in E_n and $x_o \in E_n$. Then $(S - x_o)^* = H^*(S - x_o) = C^*(S, x_o) = K^*(H(S), x_o) = H^*(K(S, x_o)) = K^*(S, x_o)$.

Proof: By Theorem 3.3.9, $C^*(S, x_o) = K^*(H(S), x_o) = H^*(K(S, x_o))$. Now since $S - x_o \subset H(S) - x_o = H(S - x_o) \subset K(H(S), x_o)$ then $(S - x_o)^* \supset H^*(S - x_o) \supset K^*(H(S), x_o)$. But note that $K^*(H(S), x_o) \supset (S - x_o)^*$. To show this let $u \in (S - x_o)^* \supset H^*(S - x_o)$ and let $y \in K(H(S), x_o)$. Then $y = \lambda(x - x_o)$ where $x \in H(S)$ and $\lambda > 0$. Noting that $\langle x - x_o, u\rangle \le 0$ then $\langle y, u\rangle = \lambda\langle x - x_o, u\rangle \le 0$. So $u \in K^*(H(S), x_o)$. The proof is complete by noting that $H^*(K(S), x_o) = K^*(S, x_o)$.

3.4. Cone of Tangents

The cones of tangents to sets at given points play an important role in developing optimality criteria. It also turns out that the cones of tangents and their polars relate to the notion of differentiability and directional differentiability of a function. In this section we intend to develop some basic results regarding these cones. These results will be used later in developing suitable optimality criteria and constraint qualifications.

3.4.1 Definition. Let S be a nonempty set in E_n and let $x_o \in C\ell\, S$. The cone of tangents to S at x_o, denoted by $T(S, x_o)$ is the set of all directions x such that $x = \lim_{k\to\infty} \lambda_k(x_k - x_o)$ where $\lambda_k > 0$, $x_k \in S$ for all k and $x_k \to x_o$.

We will interpret $T(S, x_o)$ to be the empty set if S is empty. It is obvious that $T(S, x_o)$ is a cone with vertex 0. Some examples of cones of tangents are given in Figure 3.7 below.

The following remark gives an equivalent definition of the cone of tangents. The characterization given below is especially important in revealing the relationship between the cone of tangents and some other cones to be defined later, e.g., the cone of attainable directions, the cone of interior directions, and the cone of feasible directions.

3.4.2 Remark. Let S be a nonempty set in E_n and let $x_o \in C\ell\, S$. Then $x \in T(S, x_o)$ if and only if there exists a sequence $\{x_k\}$ in S such that $x_k = x_o + \mu_k x + \mu_k \cdot \epsilon(\mu_k)$ with $\mu_k > 0$, $\mu_k \to 0$ and where ϵ is a vector function with $\epsilon(t) \to 0$ as $t \to 0$.

Proof: Let $x \in T(S, x_o)$ then $x = \lim \lambda_k(x_k - x_o)$, $\lambda_k > 0$, $x_k \in S$, i.e.,

(i) $S = \{(x,y) : y \geq -x^3\}$

$T(S,x_o) = \{(x,y) : y \geq 0\}$

where $x_o = (0,0)$.

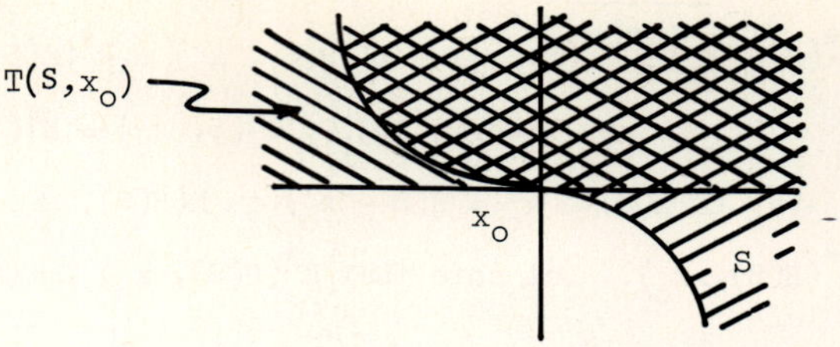

(ii) $S = \{(x,y) : x^2 + y^2 = 1\}$

$T(S,x_o) = \{(x,y) : x = 0\}$, $x_o = (-1,0)$

$T(S,x_1) = \{(x,y) : y = 0\}$, $x_1 = (0,-1)$.

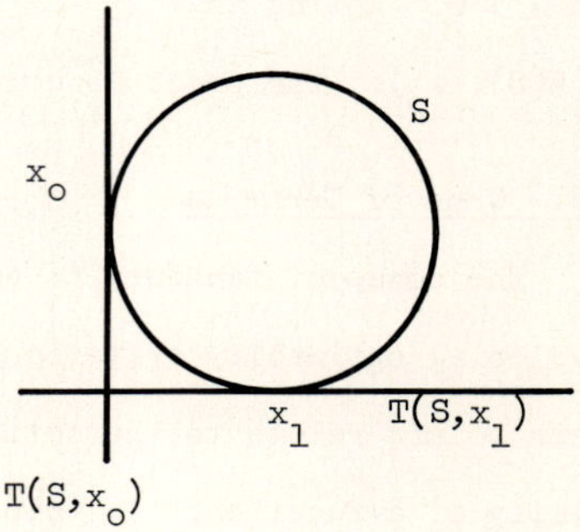

(iii) $S = \{(x,y) : x \text{ is integer}, y = 0\}$

$T(S,x_o) = \{(0,0)\}$ where $x_o = (0,0)$.

(iv) $S = \{(x,y) : x \text{ is rational}, y = 0\}$

$T(S,x_o) = E_1 x \{0\}$ where $x_o = (0,0)$.

Figure 3.7

Some Examples of Cones of Tangents

$x = \lambda_k(x_k - x_o) + \epsilon(\frac{1}{\lambda_k})$ where $\epsilon(t) \to 0$ as $t \to 0$. Obviously $\lambda_k \to \infty$ since $x_k \to x_o$ and $\lim \lambda_k(x_k - x_o)$ exists. Rearranging terms and denoting $\frac{1}{\lambda_k}$ by μ_k then $x_k = x_o + \mu_k x$ $\mu_k \cdot \epsilon(\mu_k)$. Conversely assume that there is a sequence $\{x_k\}$ in S satisfying $x_k = x_o + \mu_k x + \mu_k \cdot \epsilon(\mu_k)$ where $\mu_k \to 0$ as $k \to \infty$ and $\epsilon(t) \to 0$ as $t \to 0$. This shows that $\frac{1}{\mu_k}(x_k - x_o) \to x$ and hence $x \in T(S, x_o)$.

The reader may note from the above remark that $x \in T(S, x_o)$ if and only if given any $\delta > 0$ there is a $\lambda \in (0, \delta)$ and a point $z \in N_\lambda$ such that $x_o + \lambda x + \lambda z \in S$. N_λ is a ball about the origin with radius λ.

To summarize, given a set S and $x_o \in C\ell\, S$, then $x \in T(S, x_o)$ if and only if any of the equivalent statements hold:

(i) There exists a sequence $\{x_k\}$ and a sequence $\{\lambda_k\}$ such that $x = \lim \lambda_k(x_k - x_o)$ where $x_k \in S$, $\lambda_k > 0$ and $x_k \to x_o$. Note in this case that $\lambda_k \uparrow \infty$.

(ii) There exist a sequence $\{\mu_k\}$ and a vector function ϵ such that $x_k = x_o + \mu_k x + \mu_k \cdot \epsilon(\mu_k) \in S$ where $\mu_k \to 0$ and $\epsilon(t) \to 0$ as $t \to 0$.

(iii) Given $\delta > 0$ there is a $\lambda \in (0, \delta)$ and a $z \in N_\lambda$ such that $x_o + \lambda x + \lambda z \in S$. Here N_λ is a ball about the origin with radius $\lambda > 0$.

At this stage the reader should be able to visualize a geometrical picture of the cone of tangents. Indeed the cone of tangents is the set of directions which are tangential to the set S under consideration, i.e., a direction x belongs to $T(S, x_o)$ if there is a sequence $\{x_k\}$ in the set converging to the point x_o such that the limit of the directions given by $x_k - x_o$ is x.

The following remark follows directly from the definition of the cone of tangents.

3.4.3 <u>Remark</u>. Let S_1 and S_2 be nonempty sets in E_n and let $x_o \in C\ell\, S_1 \cap C\ell\, S_2$. Then the following statements hold:

(i) If $S_1 \subset S_2$ then $T(S_1, x_o) \subset T(S_2, x_o)$.

(ii) $T(S_1 \cap S_2, x_o) \subset T(S_1, x_o) \cap T(S_2, x_o)$.

(iii) If $x_o \in \text{int } S_2$ then $T(S_1 \cap S_2, x_o) = T(S_1, x_o)$.

<u>Proof</u>: Parts (i) and (ii) are trivial. Now we show (iii). Since $S_1 \cap S_2 \subset S_1$

then $T(S_1 \cap S_2, x_o) \subset T(S_1, x_o)$. We need to show that $T(S_1, x_o) \subset T(S_1 \cap S_2, x_o)$. Let $x \in T(S_1, x_o)$, i.e., $x = \lim \lambda_k(x_k - x_o)$ with $x_k \to x_o$, $\lambda_k > 0$, $x_k \in S_1$. But since $x_o \in \text{int } S_2$ and $x_k \to x_o$ then there is an integer K such that $x_k \in S_2$ for each $k > K$. This shows that $x \in T(S_1 \cap S_2, x_o)$ and the proof is complete.

From the above remark it is then clear that for any open ball N about x_o we must have $T(S, x_o) = T(S \cap N, x_o)$. The following theorem gives a characterization of the cone of tangents to S at $x_o \in C\ell\, S$ in terms of cones of $S \cap N$ (see Definition 3.3.1) where N is an open ball about x_o. As a corollary we show that the cone of tangents is a closed cone.

3.4.4 Theorem. Let S be a arbitrary set in E_n and $x_o \in C\ell\, S$. Then $T(S, x_o) = \bigcap_{N \in \mathcal{N}} C\ell\, K(S \cap N, x_o)$ where $\mathcal{N}$ is the class of all open balls about x_o.

Proof: If $x \in T(S, x_o)$ then $x = \lim_{k \to \infty} \lambda_k(x_k - x_o)$ with $x_k \in S$, $\lambda_k > 0$ and $x_k \to x_o$. Since $x_k \to x_o$ then given any open ball N about x_o there exists a positive integer k_K such that $x_k \in S \cap N$ for all $k > k_K$. Therefore $\lambda_k(x_k - x_o) \in K(S \cap N, x_o)$ for all $k > k_K$ and hence $x = \lim_{k \to \infty} \lambda_k(x_k - x_o) \in C\ell\, K(S \cap N, x_o)$. But since this is true for all $N \in \mathcal{N}$, then $x \in \bigcap_{N \in \mathcal{N}} C\ell\, K(S \cap N, x_o)$, i.e., $T(S, x_o) \subset \bigcap_{N \in \mathcal{N}} C\ell\, K(S \cap N, x_o)$.

Conversely let $x \in \bigcap_{N \in \mathcal{N}} C\ell\, K(S \cap N, x_o)$. Given any positive integer k choose an open ball N_k about x_o with radius 1/k. $x \in C\ell\, K(S \cap N_k, x_o)$ implies that x is the limit of vectors of the form $\lambda_\ell(x_\ell - x_o)$ where $\lambda_\ell > 0$, $x_\ell \in S \cap N_k$. Now choose ℓ_k such that $\|x - \lambda_{\ell_k}(x_{\ell_k} - x_o)\| < 1/k$. By varying k we thus generate the sequence $\{\lambda_{\ell_k}\}$ and $\{x_{\ell_k}\}$. Note that $\lambda_{\ell_k} > 0$, $x_{\ell_k} \in S$, $x_{\ell_k} \to x_o$ and $\lambda_{\ell_k}(x_{\ell_k} - x_o) \to x$, i.e., $x \in T(S, x_o)$. Therefore $T(S, x_o) \supset \bigcap_{N \in \mathcal{N}} C\ell\, K(S \cap N, x_o)$ and the proof is complete.

Corollary: $T(S, x_o)$ is a closed cone.

The fact that the cone of tangents is always closed is very important. Actually we will make use of this fact in many cases. We will next show that if S is convex at $x_o \in C\ell\, S$ then $T(S, x_o) = C\ell\, K(S, x_o)$. This result considerably simplifies evaluation of the cone of tangents for a convex set at a point. This result essentially says that we only need to consider rays from x_o through S in order to find $T(S, x_o)$. In other words it suffices to consider sequences converging to x_o through rays.

3.4.5 Remark. Let S be a set in E_n and $x_o \in C\ell\, S$. Suppose that S is convex at x_o then $T(S, x_o) = C\ell\, K(S, x_o)$.

Proof: If S is convex at x_o then from Lemma 3.3.6, $K(S, x_o) = K(S \cap N, x_o)$ for each open ball N about x_o. The result is then immediate from Theorem 3.4.4 above.

Corollary: If S is convex and $x_o \in C\ell\, S$ then $T(S, x_o)$ is a closed convex cone.

Proof: $T(S, x_o)$ is always closed. Convexity follows by Lemma 3.3.8.

From the above remark it is obvious that if a set S is convex at x_o (in particular if S is convex) then $\xi \in T^*(S, x_o)$ if and only if $\xi \in K^*(S, x_o)$. However, by Theorem 3.3.10 it then follows that $\xi \in T^*(S, x_o)$ if and only if $\langle x - x_o, \xi \rangle \leq 0$ for each $x \in S$. This fact will be used later in developing some optimality criteria of the minimum principle type.

In Lemma 3.4.3 we noted that $T(S_1 \cap S_2, x_o) \subset T(S_1, x_o) \cap T(S_2, x_o)$. However, the converse inclusion is not true in general even under convexity of S_1 and S_2. For example, let $S_1 = \{(x,y) : y \geq x^2\}$ and $S_2 = \{(x,y) : y \leq -x^2\}$ and let $x_o = (0,0)$. Note that $S_1 \cap S_2 = \{0,0\}$ and hence $T(S_1 \cap S_2, x_o) = \{(0,0)\}$ whereas $T(S_1, x_o) \cap T(S_2, x_o) = \{(x,y) : y = 0\}$. Theorem 3.4.7 below gives a sufficient condition for $T(S_1 \cap S_2, x_o) = T(S_1, x_o) \cap T(S_2, x_o)$ to hold. We will make use of this condition later to develop a qualification on the constraints of a nonlinear programming problem. To prove Theorem 3.4.7, however, we need the following lemma.

3.4.6 Lemma. Let S_1 and S_2 be nonempty convex sets in E_n and $x_o \in C\ell\, S_1 \cap C\ell\, S_2$. If $\text{ri}S_1 \cap \text{ri}S_2 \neq \emptyset$ then $\text{ri}C(S_1, x_o) \cap \text{ri}C(S_2, x_o) \neq \emptyset$.

Proof: We will show that $\text{ri}S_1 \subset x_o + \text{ri}C(S_1, x_o)$ and also $\text{ri}(S_2, x_o) \subset x_o + \text{ri}C(S_2, x_o)$. This would show that whenever $x \in \text{ri}S_1 \cap \text{ri}S_2$ then $x - x_o \in \text{ri}C(S_1, x_o) \cap \text{ri}C(S_2, x_o)$ and the result would be at hand. Note that $S_1 - x_o \subset C(S_1, x_o)$ and also note that $M(S_1 - x_o) = M(C(S_1, x_o))$ and so it then follows by Lemma 2.3.3 that $\text{ri}(S_1 - x_o) \subset \text{ri}C(S_1, x_o)$. But by the Corollary to Theorem 1.3.10 we know that $\text{ri}(S_1 - x_o) = \text{ri}S_1 - \text{ri}\{x_o\} = \text{ri}S_1 - x_o$ and therefore $\text{ri}S_1 \subset x_o + \text{ri}C(S_1, x_o)$. Similarly $\text{ri}S_2 \subset x_o + \text{ri}C(S_2, x_o)$ and the proof is complete.

3.4.7 Theorem. Let S_1 and S_2 be nonempty convex sets in E_n and $x_o \in C\ell\, S_1 \cap C\ell\, S_2$. If $\text{ri}S_1 \cap \text{ri}S_2 \neq \emptyset$ then $T(S_1 \cap S_2, x_o) = T(S_1, x_o) \cap T(S_2, x_o)$.

Proof: To prove the Theorem, in view of Lemma 3.4.5, we need to show that $C\ell\, C(S_1 \cap S_2, x_o) = C\ell\, C(S_1, x_o) \cap C\ell\, C(S_2, x_o)$. Here $C(S, x_o)$ and $C(S_2, x_o)$ are the convex cones of S_1 and S_2 at x_o. It is obvious that $C(S_1 \cap S_2, x_o) \subset C(S_1, x_o) \cap$

$C(S_2, x_o)$ and hence $C\ell\, C(S_1 \cap S_2, x_o) \subset C\ell\, (C(S_1, x_o) \cap C(S_2, x_o)) \subset C\ell\, C(S_1, x_o) \cap C\ell\, C(S_2, x_o)$. To show the converse inclusion let $x \in C\ell\, C(S_1, x_o) \cap C\ell\, C(S_2, x_o)$ and let $y \in riC(S_1, x_o) \cap riC(S_2, x_o)$ which exists by Lemma 3.4.6. Consider $x_\lambda = \lambda y + (1 - \lambda)x$ for $\lambda \in (0,1)$. By Theorem 2.3.5 it follows that $x_\lambda \in riC(S_1, x_o) \cap riC(S_2, x_o) \subset C(S_1, x_o) \cap C(S_2, x_o)$ for each $\lambda \in (0,1)$. But by Lemma 3.3.5 we have $C(S_1, x_o) \cap C(S_2, x_o) = C(S_1 \cap S_2, x_o)$ and so $x_\lambda \in C(S_1 \cap S_2, x_o)$ for each $\lambda \in (0,1)$. Therefore $x = \lim_{\lambda \to 0} x_\lambda \in C\ell\, C(S_1 \cap S_2, x_o)$ and the proof is complete.

We will now develop the relationship between the cone of tangents of a set S and the cone of tangents of its complement S^c. For this purpose we need the following lemmas.

3.4.8 <u>Lemma</u>. Let S be a nonempty set in E_n and $x_o \in C\ell\, S$. Then $T(S, x_o) \cup T(S^c, x_o) = E_n$.

<u>Proof</u>: Let $x \in E_n$ and consider the sequence $\{x_k\}$ given by $x_k = x_o + \frac{1}{k} x$. Obviously $x_k \to x_o$ as $k \to \infty$. If $\{x_k\}$ has a subsequence in S then clearly $x \in T(S, x_o)$ by letting $\lambda_k = k$ in the definition of the cone of tangents. On the other hand if no such subsequence exists then there is a positive integer N such that $x_{N+i} \in S^c$ for all $i > 0$. It then follows that $x \in T(S^c, x_o)$, i.e., $x \in T(S, x_o) \cup T(S^c, x_o)$. This completes the proof.

3.4.9 <u>Lemma</u>. Let S_1 and S_2 be nonempty subsets of E_n such that $S_1 \cup S_2 = E_n$. Suppose there exists a nonzero $p_1 \in S_1^*$ and a nonzero $p_2 \in S_2^*$. Let H_1 and H_2 be the hyperplanes through the origin with normals p_1 and p_2, i.e., $H_1 = \{x : \langle x, p_1 \rangle = 0\}$ and $H_2 = \{x : \langle x, p_2 \rangle = 0\}$. Then the following statements hold.

(i) $C\ell\, S_1 \subset H_1^-$ and $C\ell\, S_2 \subset H_2^-$.

(ii) $p_1 = \lambda p_2$ for some $\lambda < 0$.

(iii) $C\ell\, S_1 \cap C\ell\, S_2 = H_1 = H_2$.

<u>Proof</u>: $p_1 \in S_1^*$ implies that $S_1 \subset H_1^-$. Likewise $p_2 \in S_2^*$ implies that $S_2 \subset H_2^-$ and hence (i) holds. Also $H_1^- \cup H_2^- = E_n$. Now let $x \in H_1$. If $\langle x, p_2 \rangle > 0$ then there is an open ball N about x such that $N \subset (H_2^-)^c$, the complement of H_2^-. But since x is on the boundary of H_1^- then there is a point $x_1 \in (H_1^-)^c \cap N$, i.e., $x_1 \in (H_1^-)^c \cap (H_2^-)^c$, which is impossible since $H_1^- \cup H_2^- = E_n$. If $\langle x, p_2 \rangle < 0$ then the same argument applies for $-x$ and hence we conclude that $\langle x, p_2 \rangle = 0$. This implies that $x \in H_2$ and

so $H_1 \subset H_2$. By symmetry $H_1 = H_2$ and hence there is a $\lambda < 0$ such that $p_1 = \lambda p_2$, and so (ii) holds. Now if $x \in C\ell\, S_1 \cap C\ell\, S_2$ then $\langle x, p_1\rangle \leq 0$ and $\langle x, p_2\rangle \leq 0$ and since $p_1 = \lambda p_2$ with $\lambda < 0$ we conclude that $\langle x, p_1\rangle = \langle x, p_2\rangle = 0$, i.e., $x \in H_1$ and (iii) then follows.

The following theorem shows that if the polars of the cones of tangents to a set and to its complement are both nonzero then the cones of tangents are halfspaces and the polar cones are rays through the origin. In some sense this shows that the set underconsideration is "smooth" at the point under consideration. Indeed, if the set under consideration is the set above a given function, then the cone of tangents to the set corresponding to a point where the function is differentiable is a halfspace. This is due to the "smoothness" property which is in turn due to differentiability. This matter will be discussed in the example following Theorem 3.4.10 and will be discussed in detail in Chapter 6. Figure 3.8 shows an example where the cones of tangents are halfspaces with normals ξ and $-\xi$ and another example where this is not the case. In this latter situation the polar cone $T(S^c, x_o)$ is the zero vector. From Lemma 3.4.8 and 3.4.9, the following readily follows.

3.4.10 Theorem. Let S be a nonempty set in E_n and $x_o \in C\ell\, S$. If $T^*(S, x_o) \neq \{0\}$ and $T^*(S^c, x_o) \neq \{0\}$ then there exists a nonzero vector p such that $T(S, x_o) = \{x : \langle x, p\rangle \leq 0\}$ and $T(S^c, x_o) = \{x : \langle x, p\rangle \geq 0\}$. Moreover $T^*(S, x_o) = \{\lambda p : \lambda \geq 0\}$ and $T^*(S^c, x_o) = \{\lambda p : \lambda \leq 0\}$.

Now consider the function $f(x) = x^2 + x$ and the set $S = \{x : x^2 + x \leq 0\}$. Let $x_o = 0$. Let $\xi \in T(S, x_o)$ and denote the gradient vector of f at x_o by $\nabla f(x_o) = 1$. We will show that $\nabla f(x_o) \in T^*(S, x_o)$ by showing that $\langle \xi, \nabla f(x_o)\rangle \leq 0$. Note that $\xi = \lim \lambda_k(x_k - x_o)$ where $\lambda_k > 0$, $x_k \to x_o$, and $f(x_k) \leq f(x_o) = 0$. Now $f(x_k) = f(x_o) + \langle x_k - x_o, \nabla f(x_o)\rangle + \|x_k - x_o\|\, \epsilon(x_k - x_o)$. Note that $f(x_o) = 0$, $f(x_k) \leq 0$ and $\epsilon(t) \to 0$ as $t \to 0$. Therefore, multiplying by λ_k and letting $k\to\infty$ we get $\langle \xi, \nabla f(x_o)\rangle \leq 0$. Therefore, $\nabla f(x_o) \in T^*(S, x_o)$. Similarly it can be easily shown that $-\nabla f(x_o) \in T^*(S^c, x_o)$. But since $\nabla f(x_o) = 1$ then we have shown that $T^*(S, x_o)$ and $T^*(S^c, x_o)$ are nonzero. Therefore, from Theorem 3.4.10 it follows that $T(S, x_o)$ and $T(S^c, x_o)$ are halfspaces given by the nonpositive and the nonnegative reals respectively. This is illustrated in Figure 3.9.

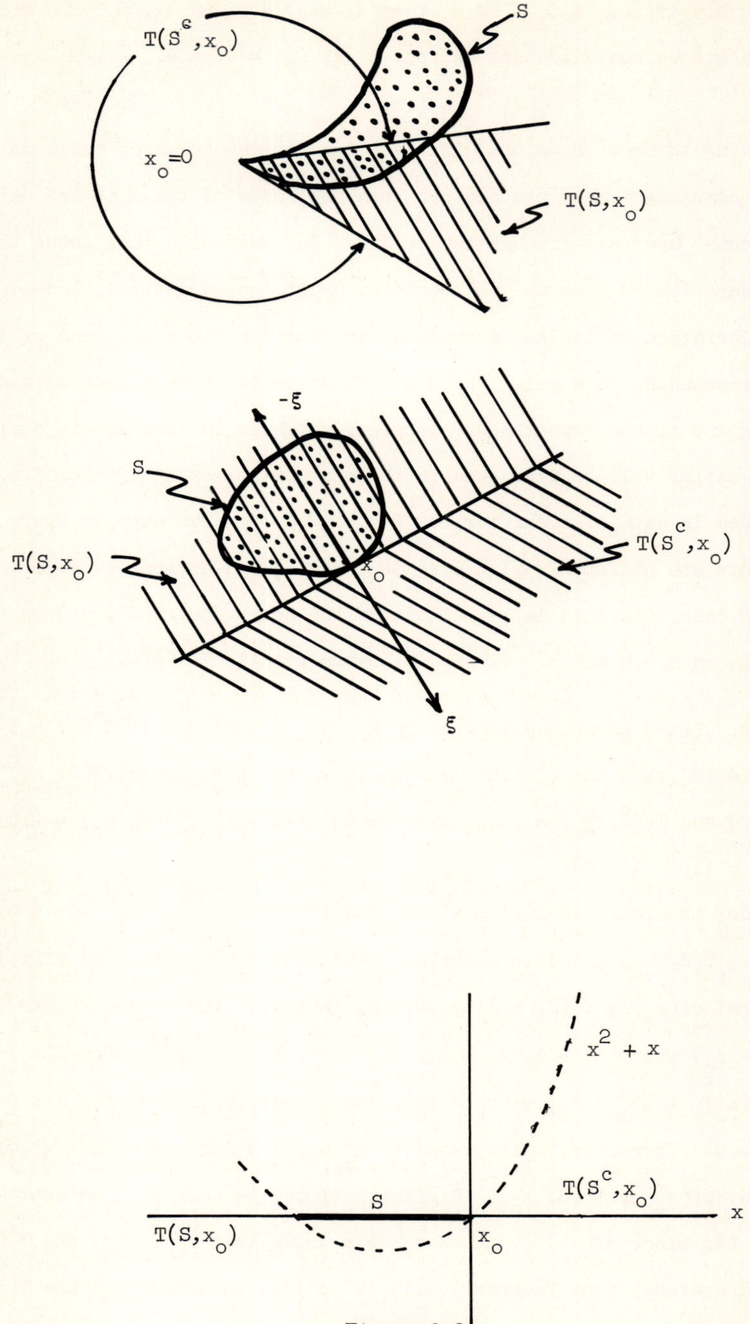

Figure 3.9

Example of the Cone of Tangents

3.5 Cone of Attainable Directions, Cone of Feasible Directions, and Cone of Interior Directions

The cone of attainable directions, the cone of feasible directions, and the cone of interior directions represent, in some sense, further restriction of the cone of tangents. These cones are important from the point of view of nonlinear programming and they play a significant role in developing optimality criteria and various constraint qualifications. We will first consider the cone of attainable directions. This cone is very similar to the cone of tangents except for the fact that the sequence of points $\{x_n\}$ in Definition 3.4.1 is replaced by a continuous sequence or a feasible arc.

3.5.1 Definition. Let S be a nonempty set in E_n and $x_o \in C\ell\, S$. x is said to be in the cone of attainable directions of S at x_o, denoted by $A(S, x_o)$ if there exists a $\delta > 0$ and a vector valued function α: $E_1^+ \to E_n$ which defines an arc such that $\alpha(0) = x_o$, $\alpha(\theta) \in S$ for each $\theta \in (0,\delta]$ and $\left.\frac{d\alpha(\theta)}{d\theta}\right|_{\theta=0} = x$.

3.5.2 Lemma. Let S be a nonempty set in E_n and $x_o \in C\ell\, S$. Then $x \in A(S, x_o)$ if and only if there exists a function $\alpha : E_1^+ \to E_n$ such that $\alpha(\theta) = x_o + \theta x + \theta \cdot \epsilon(\theta)$ belongs to S for each $\theta \in (0,\delta]$ and $\epsilon(\theta) \to 0$ as $\theta \to 0$.

Proof: If $x \in A(S, x_o)$ then there exists a function α satisfying Definition 3.5.1. But $\left.\frac{d\alpha(\theta)}{d\theta}\right|_{\theta=0} = x$ implies that $\lim_{\theta \to 0+} \frac{\alpha(\theta) - x_o}{\theta} = x$ which in turn implies that $\alpha(\theta) - x_o = \theta x + \theta \cdot \epsilon(\theta)$ where $\epsilon(\theta) \to 0$ as $\theta \to 0$. In other words $\alpha(\theta) = x_o + \theta x + \theta \cdot \epsilon(\theta)$ with $\alpha(\theta) \in S$ for each $\theta \in (0,\delta]$ where δ is given according to Definition 3.5.1. Conversely suppose that an arc $\alpha(\theta) = x_o + \theta x + \theta \cdot \epsilon(\theta)$ exists with the specified properties. Then $x_o = \alpha(0)$ and $\lim_{\theta \to 0+} \frac{\alpha(\theta) - x_o}{\theta} = x + \lim_{\theta \to 0} \epsilon(\theta) = x$, i.e., $x \in A(S, x_o)$.

From this remark it is evident that the cone of attainable directions is a stronger version of the cone of tangents where we require an arc rather than a sequence of points. In some sense we need a "connected" sequence or "continuous" sequence of points rather than the "discrete" points required for the cone of tangents. Therefore it is always true that $A(S, x_o) \subset T(S, x_o)$.

Figure 3.10 gives some examples of the cones of attainable directions. It should be noted that the cone of attainable directions is not necessarily nonempty as

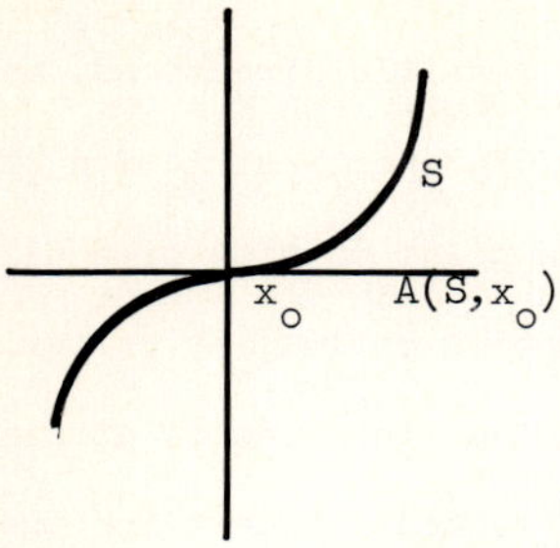

(i) $S = \{(x,y) : y = x^3\}$

$A(S, x_o) = \{(x,y) : y = 0\}$

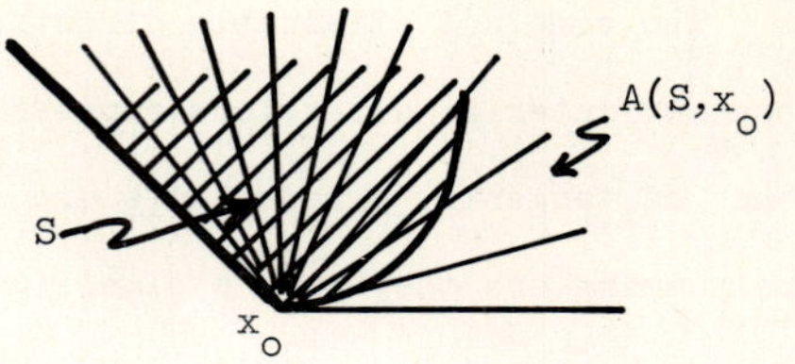

(ii) $S = \{(x,y) : x \geq 0,\ y \geq x^3\} \cup$

$\{(x,y) : x \leq 0,\ y \geq -x\}$

$A(S, x_o) = \{(x,y) : y \geq \max\ (0, -x)$

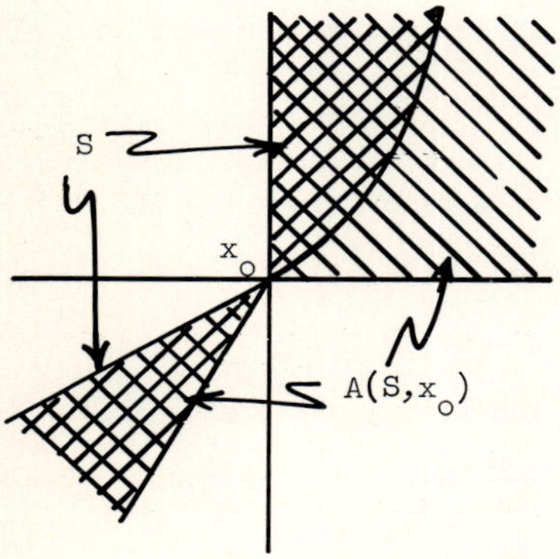

$S = \{(x,y) : x \geq 0,\ y \geq x^2\} \cup$

$\{(x,y) : x \leq 0\ -\frac{2}{3}x \leq y \leq -\frac{1}{3}x\}$

(iii) $A(S,x_o) = \{(x,y) : x \geq 0,\ y \geq 0\} \cup$

$\{(x,y) : x \leq 0\ -\frac{2}{3}x \leq y \leq -\frac{1}{3}x\}$

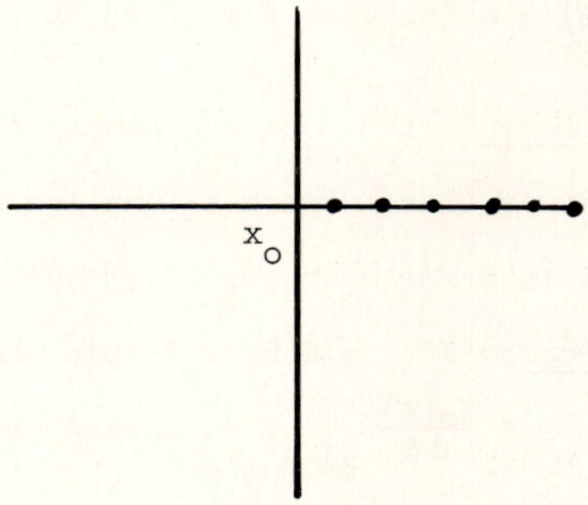

(iv) $S = \{(x,y) :$ x is rational,

$y = 0\} \sim \{0,0\}$

$A(S, x_o) = \emptyset$ where $x_o = (0,0)$

Figure 3.10

Some Examples of Cones of Attainable Directions

was the case for the cone of tangents. As an example, Figure 3.12 (iv) shows a case where $A(S, x_o) = \emptyset$ since there are no arc in S. However, $T(S, x_o) = \{(x,y) : x \geq 0\}$.

We may mention that some of the properties mentioned earlier for the cones of tangents hold for the cones of attainable directions, but not all. The following remark is a parallel of Lemma 3.4.3 for the cones of tangents. The proof is also similar and hence is omitted.

3.5.3 Lemma. Let S_1 and S_2 be nonempty sets in E_n and let $x_o \in C\ell\, S_1 \cap C\ell\, S_2$. Then the following statements hold.

(i) If $S_1 \subset S_2$ then $A(S_1, x_o) \subset A(S_2, x_o)$.

(ii) $A(S_1 \cap S_2, x_o) \subset A(S_1, x_o) \cap A(S_2, x_o)$.

(iii) If $x_o \in \text{int } S_2$ then $A(S_1 \cap S_2, x_o) = A(S_1, x_o)$.

However, it is worthwhile mentioning that some of the important characteristics of the cone of tangents do not hold in the case of the cone of attainable directions. Also $A(S, x_o) \cup A(S^c, x_o) \neq E_n$ in general. For example, let S be the set of points in E_2 with rational coordinates. Then $A(S, x_o) = \{0\}$ and $A(S^c, x_o) = \emptyset$. Hence $A(S, x_o) \cup A(S^c, x_o) = \{0\}$.

We will now show that if the set S is convex at x_o where $x_o \in C\ell\, S$ then the cones of tangents and the cone of attainable directions are essentially equal.

3.5.4 Lemma. Let S be a nonempty set and $x_o \in C\ell\, S$. Further suppose that S is convex at x_o. Then $C\ell\, A(S, x_o) = T(S, x_o)$.

Proof: We will first show that any nonzero $x \in K(S, x_o)$ is also in $A(S, x_o)$. But $x \in K(S, x_o)$ means that $x = \lambda(y - x_o)$ where $y \in S$ and $\lambda > 0$. Now consider $\alpha(\theta) = x_o + \theta x = x_o + \theta\lambda y - \theta\lambda x_o = \theta\lambda y + (1 - \theta\lambda)x_o$. Therefore, for $\theta \in \left(0, \frac{1}{\lambda}\right)$ we have $\theta\lambda \in (0,1)$ and $\alpha(\theta) \in S$ since S is convex at x_o (see Definition 3.3.4). This means that $x \in A(S, x_o)$ according to Lemma 3.5.2, where ε is the zero function. This shows that $C\ell\, K(S, x_o) \subset C\ell\, A(S, x_o)$. But by Lemma 3.4.5 it follows that $T(S, x_o) = C\ell\, K(S, x_o)$ and therefore $T(S, x_o) \subset C\ell\, A(S, x_o)$. But on the other hand $A(S, x_o) \subset T(S, x_o)$ as discussed earlier (this is trivial to check) and by the fact that $T(S, x_o)$ is closed it follows that $C\ell\, A(S, x_o) \subset T(S, x_o)$, i.e., $T(S, x_o) = C\ell\, A(S, x_o)$.

Corollary 1: Let S_1 and S_2 be convex sets with $x_o \in C\ell\, S_1 \cap C\ell\, S_2$ and $\text{ri} S_1 \cap \text{ri} S_2 \neq \emptyset$. Then, $C\ell\, A(S_1 \cap S_2, x_o) = C\ell\, A(S_1, x_o) \cap C\ell\, A(S_2, x_o)$.

Proof: This is just a restatement of Theorem 3.4.7 in view of the above result.

Corollary 2: If S is convex then $C\ell\, A(S, x_o)$ is convex.

In some sense, Corollary 1 above gives the converse inclusion of Lemma 3.4.2 part (ii) above. Corollary 2 shows that $C\ell\, A(S, x_o)$ is convex as long as S is convex. Actually one can say more than that. Indeed if S is convex then $A(S, x_o)$ is convex. This can be directly checked and is left as an exercise for the reader.

We will now discuss a special case of the cone of attainable directions, namely the cone of feasible directions. In this case we insist that the arc contained in the set be "linear." A precise definition is given below.

3.5.5 Definition. Let S be a nonempty set in E_n and let $x_o \in C\ell\, S$. Then x is said to be in the cone of feasible directions to S at x_o, denoted by $F(S, x_o)$ if there exists a $\delta > 0$ such that $x_\lambda = x_o + \lambda x$ is in S for each $\lambda \in (0, \delta]$.

From this definition it is clear that $F(S, x_o)$ is a special case of $A(S, x_o)$ where ϵ is identically zero. In other words the arc in the cone of attainable directions is a linear arc or a ray going from the point x_o. Of course, this requirement simplifies a lot of problems associated with nonlinearities in the general arcs. It will be evident in later chapters that it is appropriate to work with the cone of feasible directions (and cones of interior directions to be discussed below) for sets of the inequality type, and to work with the cone of attainable directions or the cone of tangents for sets of the equality type.

We give some examples of cones of feasible directions in Figure 3.12. Note that $F(S, x_o)$ may be empty as in Figure 3.12(a) if we exclude the origin from S. Also note that the cone may or may not be either closed or open. (see Figure 3.12(b) and (d) respectively). Also note that $F(S, x_o) \cup F(S^c, x_o)$ is not E_n. For example, let S be the set of rational real numbers and x_o be the origin. Then $F(S, x_o) = \{0\}$ and $F(S^c, x_o) = \emptyset$. Hence $F(S, x_o) \cup F(S^c, x_o) = \{0\}$.

From the above it is evident that $F(S, x_o) \subset A(S, x_o) \subset T(S, x_o)$. It is also clear from Lemma 3.5.4 that if S is convex at x_o then $C\ell\, F(S, x_o) = C\ell\, A(S, x_o) = T(S, x_o)$. Note that the same proof as in Lemma 3.5.4 will hold if $A(S, x_o)$ is replaced by $F(S, x_o)$.

We found out earlier that in general $T(S_1, x_o) \cap T(S_2, x_o) \neq T(S_1 \cap S_2, x_o)$.

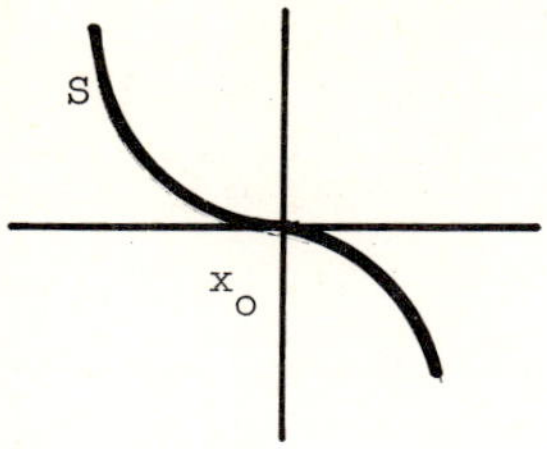

(a) $S = \{(x,y) : y \{ -x^3\}$

$F(S, (0,0)) = \{0\}$

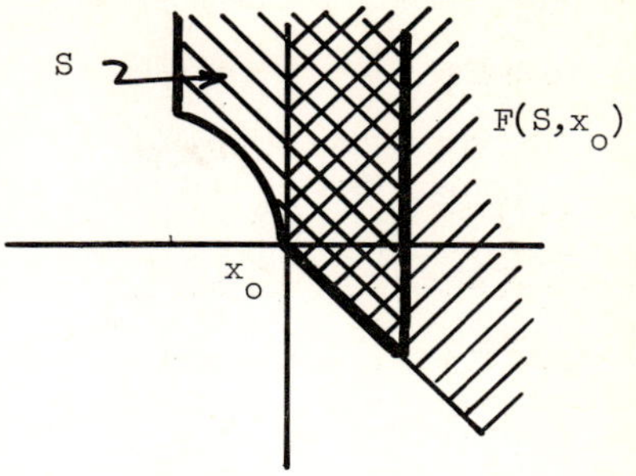

(b) $S = \{(x,y) : -1 \le x \le 1,\ y \ge x^{1/3},\ y \ge -x\}$

$F(S, (0,0)) = \{(x,y) : x \ge 0,\ y \ge x\}$.

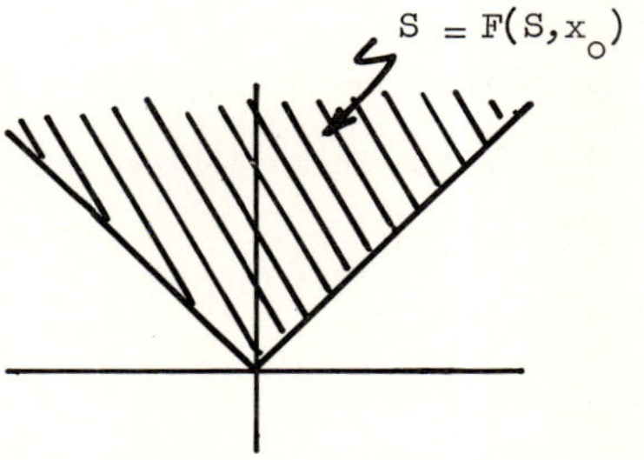

(c) $S = \{(x,y) : y \ge |x|\}$

$F(S, (0,0)) = \{(x,y) : y \ge |x|\}$

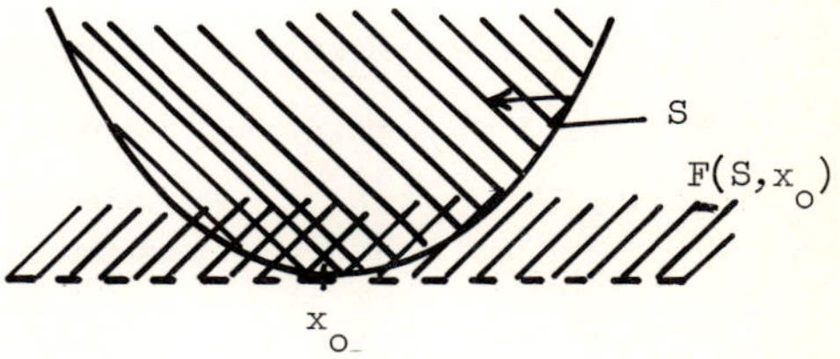

(d) $S = \{(x,y) : y > x^2\}$

$F(S, (0,0)) = \{(x,y) : y > 0\}$

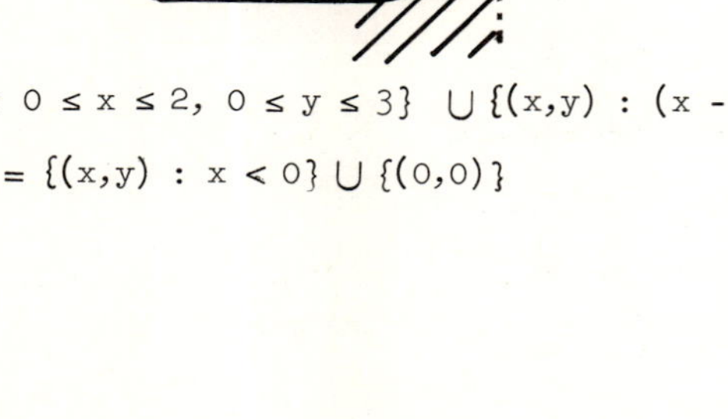

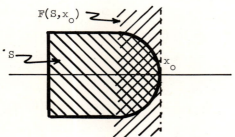

(e) $S = \{(x,y) : 0 \le x \le 2,\ 0 \le y \le 3\} \cup \{(x,y) : (x - 2)^2 + (y - 1)^2 \le 1\}$

$F(S, (3,1)) = \{(x,y) : x < 0\} \cup \{(0,0)\}$

Figure 3.12

Some Examples of Cones of Feasible Directions

However, this important property holds true for the cones of feasible directions as proved below.

3.5.6 Lemma. Let S_1 and S_2 be nonempty sets in E_n and $x_o \in C\ell\, S_1 \cap C\ell\, S_2$. Then $F(S_1 \cap S_2, x_o) = F(S_1, x_o) \cap F(S_2, x_o)$.

Proof: Let $x \in F(S_1 \cap S_2, x_o)$. Then there exists a $\delta > 0$ such that $x_\lambda = x_o + \lambda x \in S_1 \cap S_2$ for each $\lambda \in (0, \delta]$. But this automatically implies that $x \in F(S_1, x_o) \cap F(S_2, x_o)$. Conversely let $x \in F(S_1, x_o) \cap F(S_2, x_o)$ then there exists $\delta_1, \delta_2 > 0$ such that $x_\lambda = x_o + \lambda x \in S_1$ for each $\lambda \in (0, \delta_1]$ and $x_\lambda = x_o + \lambda x \in S_2$ for each $\lambda \in (0, \delta_2]$. Letting $\delta = \min(\delta_1, \delta_2)$ then $x_\lambda \in S_1 \cap S_2$ for each $\lambda \in (0, \delta)$ and so $x \in F(S_1 \cap S_2, x_o)$. This completes the proof.

Another cone which will be of importance to us is the cone of interior directions to a set S at a point x_o, denoted by $I(S, x_o)$. This cone is defined below.

3.5.7 Definition. Let S be a nonempty set in E_n and $x_o \in C\ell\, S$. The cone of interior directions to S at x_o, denoted by $I(S, x_o)$ is given by $I(S, x_o) = \{x$: there is a ball N about the origin and a $\delta > 0$ such that $y \in x + N$ and $\lambda \in (0, \delta)$ imply that $x_o + \lambda y \in S\}$.

From the above definition it is clear that $I(S, x_o)$ is indeed an open cone. This property of being an open set will be useful in developing certain optimality criteria in later chapters. From the definition it is clear that x belongs to $I(S, x_o)$ if there is a neighborhood about x and a $\delta > 0$ such that for any y in this neighborhood, $x_o + \lambda y$ belongs to the set S for each $\lambda \in (0, \delta)$. This shows that $I(S, x_o) \subset F(S, x_o)$ and since $I(S, x_o)$ is always open then $I(S, x_o) \subset \text{int } F(S, x_o)$.

The reader may be tempted to believe that the other inclusion holds, i.e., $I(S, x_o) = \text{int } F(S, x_o)$. However, this is true in general. The intuitive reason is that if $x \in \text{int } F(S, x_o)$ then there is a neighborhood about x such that any y in this neighborhood belongs to $F(S, x_o)$. So for each such y there is a $\delta_y > 0$ such that $x_o + \lambda y \in S$ for each $\lambda \in (0, \delta_y)$. Note that for the case of $I(S, x_o)$ the δ is fixed but in the case of the interior of $F(S, x_o)$ the δ is depending upon y and it may be the case that $\inf \delta_y = 0$. As an example where $\text{int } F(S, x_o) \neq I(S, x_o)$, consider $S = \{(x,y) : y \geq x^2\} \cup \{(x,y) : y \leq 0\}$ and $x_o = (0,0)$. It can be easily shown that

$F(S, x_o) = E_2$ and so int $F(S, x_o) = E_2$. On the other hand $I(S, x_o) = \{(x,y) : y > 0\} \cup \{(x,y) : y < 0\}$. In other words the cone of interior directions does not contain the x axis.

The following lemma summarizes some important properties about the cone of interior directions.

3.5.8 Lemma. Let S be a nonempty set in E_n and $x_o \in C\ell\, S$. Then

(i) $I(S, x_o)$ is an open cone.

(ii) If S is convex then $I(S, x_o)$ is convex.

If S_1 and S_2 are nonempty sets in E_n with $x_o \in C\ell\, S_1 \cap C\ell\, S_2$ then $I(S_1, x_o) \cap I(S_2, x_o) = I(S_1 \cap S_2, x_o)$.

CHAPTER 4

CONVEX FUNCTIONS

In this chapter we deal with various properties of convex functions. Continuity, directional differentiability, and supportability properties of convex functions are discussed. Also different examples of convex functions are given. Both weaker and stronger forms of convexity are also given.

4.1. Definitions and Preliminary Results

4.1.1 Definition. Let $f : S \to E_1 \cup \{\infty\}$ where S is a convex set. f is said to be convex on S (or simply convex) if $f(\lambda x + (1 - \lambda)y) \leq \lambda f(x) + (1 - \lambda)f(y)$ for each $x,y \in S$ and each $\lambda \in (0,1)$. f is said to be strictly convex if the strict inequality above holds for each distinct $x,y \in S$ with $f(x)$ and $f(y) < \infty$ and where $\lambda \in (0,1)$. The Effective Domain (or simply domain) of f is $\{x \in S: f(x) < \infty\}$ and is denoted by dom f.

One can define concave function in a similar fashion. Actually a function g is concave on S if -g is convex on S. The reader can easily check that this is equivalent to $g(\lambda x + (1 - \lambda)y) \geq \lambda g(x) + (1-\lambda)g(y)$ for each $x,y \in S$ and each $\lambda \in (0,1)$. Strict concavity is similarly defined. Finally dom $g = \{x \in S : g(x) > -\infty\}$. Obviously strict convexity implies convexity and strict concavity implies concavity.

One can equivalently define convex and concave functions via the epigraph and hypograph of the function defined below.

4.1.2 Definition. Let $f : S \to E_1 \cup \{\infty\}$. The epigraph of f, denoted by E_f is given by $E_f = \{(x,y) \in E_n \times E_1 : x \in S,\ y \in E_1,\ y \geq f(x)\}$. Let $g : S \to E_1 \cup \{-\infty\}$. The hypograph of g, denoted by H_g is given by $H_g = \{(x,y) \in E_n x E_1 : x \in S,\ y \in E_1,\ y \leq g(x)\}$.

The reader may note carefully that $(x,y) \in E_f$ automatically implies that y is a real number. In other words $y \neq \infty$. Similarly $(x,y) \in H_g$ implies that $y \neq -\infty$. The following figure shows an example of a convex function and its epigraph as well as an example of a concave function and its hypograph. Note that f is not continuous at x_o and E_f is not closed.

Now suppose we are given a convex function which is defined on a proper subset S

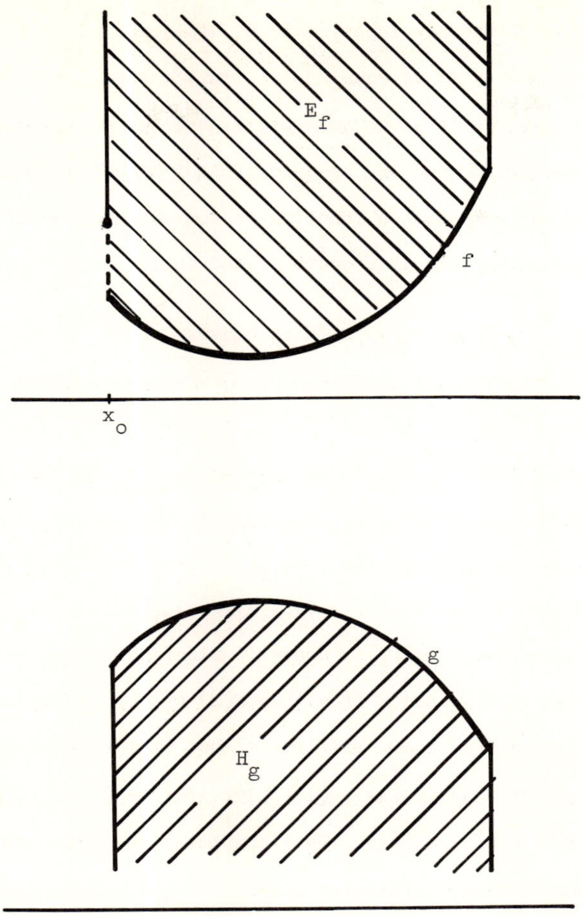

Figure 4.1

Examples of Convex and Concave Functions

With Illustration of the Epigraph and Hypograph

of E_n. It is clear that we can extend f to a convex function which is defined on E_n by letting the value of the new function be ∞ outside S. Clearly the restriction of the new function to the set S is equal to f. This is sometimes helpful in the presentation. In other words if we do not specify the set on which a convex (concave) function is defined, then we are assuming that f is defined everywhere in E_n.

The following lemma shows that a function is convex if and only if its epigraph is a convex set. Also a function is concave if and only if its hypograph is convex.

4.1.3 Lemma. Let $f : S \to E_1 \cup \{\infty\}$ where S is convex. f is convex on S if and only if E_f is convex. Similarly, let $g : S \to E_1 \cup \{-\infty\}$ where S is convex. g is concave on S if and only if H_g is convex.

Proof: Now suppose f is convex, i.e., $f(\lambda x+(1-\lambda)y) \leq \lambda f(x) + (1-\lambda)f(y)$ for each x, $y \in S$ and $\lambda \in (0,1)$. Let (x_1,μ_1), $(x_2,\mu_2) \in E_f$, i.e., $\mu_1 \geq f(x_1)$ and $\mu_2 \geq f(x_2)$. Then $\lambda\mu_1 + (1 - \lambda)\mu_2 \geq \lambda f(x_1) + (1 - \lambda)f(x_2) \geq f(\lambda x_1 + (1 - \lambda)x_2)$ for each $\lambda \in (0,1)$. This implies that $(\lambda x_1 + (1 - \lambda)x_2, \lambda\mu_1 + (1 - \lambda)\mu_2) \in E_f$, i.e., E_f is convex. Conversely assume that E_f is convex. Clearly the inequality $f(\lambda x + (1 - \lambda)y) \leq \lambda f(x) + (1 - \lambda)f(y)$ holds if either f(x) or f(y) is infinite. So suppose that both f(x) and $f(y) < \infty$. This means that $(x,f(x))$ and $(y,f(y)) \in E_f$ and by convexity of the latter we have $(\lambda x + (1 - \lambda)y, \lambda f(x) + (1 - \lambda)f(y)) \in E_f$, i.e., $f(\lambda x + (1 - \lambda)y) \leq \lambda f(x) + (1 - \lambda)f(y)$. This completes the proof for the convex case. The concave case can be similarly proved.

The notion of convexity implies some interesting properties. These are summarized below and can easily be proved.

4.1.4 Lemma. Let $f : S \to E_1 \cup \{\infty\}$ where S is a convex set in E_n. Then dom f is convex. Also $\{x : f(x) \leq \alpha\}$ is convex for each $\alpha \in E_1$. Similarly, let $g : S \to E_1 \cup \{-\infty\}$ where S is a convex set in E_n. Then dom g is convex. Also $\{x : g(x) \geq \alpha\}$ is convex for each $\alpha \in E_1$.

The converse of the above statement may not hold in general. For example consider $f(x) = x^3$. Clearly $\{x : f(x) \leq \alpha\}$ is convex for each real α. On the other hand f is not convex.

The following theorem shows that the epigraph of a convex function f (which is a convex set) has a nonvertical supporting hyperplane at a point $(x_0,f(x_0))$ where

$x_o \in$ ri dom f. The proof is hinged around the fact that a convex set (the epigraph of f) has a proper supporting hyperplane at each point in the relative boundary. Clearly $(x_o, f(x_o)) \in$ rb E_f.

4.1.5 Theorem. Let $f : S \to E_1 \cup \{\infty\}$ be a convex function on S where S is a convex set. Let $x_o \in$ ri dom f. Then there is a supporting hyperplane to E_f at $(x_o, f(x_o))$ such that $\langle (x,y) - (x_o, f(x_o)), (\xi, -1) \rangle \le 0$ for each $(x,y) \in E_f$ where $\xi \in E_n$.

Proof: E_f is a convex set and $(x_o, f(x_o)) \in$ rb E_f. Then by Theorem 2.4.9 there is a hyperplane H that supports E_f properly, i.e., there is a nonzero $(\hat{\xi}, \mu) \in E_{n+1}$ such that $\langle (x,y) - (x_o, f(x_o)), (\hat{\xi}, \mu) \rangle \le 0$ for each $(x,y) \in E_f$ and E_f is not contained in the hyperplane $H = \{(x,y) : \langle (x,y) - (x_o, f(x_o)), (\hat{\xi}, \mu) \rangle = 0\}$. Clearly μ cannot be positive because the above inequality will be violated by choosing y sufficiently large. Now we will show that μ cannot be zero. By contradiction, suppose that $\mu = 0$. Then $\langle x - x_o, \hat{\xi} \rangle \le 0$ for each $x \in$ dom f. But since $x_o \in$ ri dom f and the hyperplane $\{x : \langle x - x_o, \hat{\xi} \rangle = 0\}$ supports dom f at x_o, then by Theorem 2.4.10 we must have $\langle x - x_o, \hat{\xi} \rangle = 0$ for each $x \in$ dom f. Combining this with the assumption that $\mu = 0$ we conclude that the hyperplane H does not support E_f properly at $(x_o, f(x_o))$, a contradiction. This then shows that $\mu < 0$. Dividing by $-\mu$ and letting $\xi = \hat{\xi}/-\mu$, the result is immediate.

We have seen from the above theorem that the epigraph of a convex function has a nonvertical supporting hyperplane for points corresponding to points in the relative interior of the domain of the function. Actually these supporting hyperplanes correspond to subgradients of convex functions. The following definition gives the notion of a subgradient. This notion will be generalized to include nonconvex (and nonconcave) functions.

4.1.6 Definition. Let $f : S \to E_1 \cup \{\infty\}$ be a convex function on the convex set S. Then ξ is said to be a subgradient of f at $x_o \in S$ if $f(x) \ge f(x_o) + \langle x - x_o, \xi \rangle$ for each $x \in S$. Let $g : S \to E_1 \cup \{-\infty\}$ be a concave function on the convex set S. Then ξ is said to be a subgradient of g at $x_o \in S$ if $g(x) \le g(x_o) + \langle x - x_o, \xi \rangle$ for each $x \in S$.

The following theorem follows from Theorem 4.1.5 and the above definition. It essentially shows that a convex function has a subgradient at each point in the relative interior of its domain.

4.1.7 Theorem. Let S be a convex in E_n and $f : S \to E_1 \cup \{\infty\}$. Then f is convex on S only if for each $x_o \in$ ri dom f there exists a vector $\xi \in E_n$ such that $f(x) \geq f(x_o) + \langle x - x_o, \xi\rangle$ for every $x \in S$. Conversely, if for each $x_o \in$ ri dom f there is a vector ξ with $f(x) \geq f(x_o) + \langle x - x_o, \xi\rangle$ for each $x \in S$ then f is convex on ri dom f.

Proof: First suppose that f is convex on S. Then by Theorem 4.1.5 there exists an $\xi \in E_n$ such that $\langle x - x_o, \xi\rangle - (y - f(x_o)) \leq 0$ for each $(x,y) \in E_f$, i.e., for each $x \in$ dom f and $y \geq f(x)$. Therefore $f(x) \geq f(x_o) + \langle x - x_o, \xi\rangle$ for each $x \in$ dom f. Since $f(x) = \infty$ for $x \in S \sim$ dom f then clearly $f(x) \geq f(x_o) + \langle x - x_o, \xi\rangle$ for every $x \in S$. Conversely, let $x_1, x_2 \in$ ri dom f. By convexity of dom f (and hence convexity of ri dom f) then we must have $\lambda x + (1 - \lambda)x_2 \in$ ri dom f for each $\lambda \in (0,1)$. Therefore by the hypothesis of the theorem it follows that

$$f(x_1) \geq f(\lambda x_1 + (1 - \lambda)x_2) + \langle (1 - \lambda)(x_1, x_2), \xi\rangle , \quad \text{and}$$

$$f(x_2) \geq f(\lambda x_1 + (1 - \lambda)x_2) + \langle \lambda(x_2 - x_1), \xi\rangle$$

where ξ is a subgradient of f at $\lambda x_1 + (1 - \lambda)x_2$. Multiplying the above two inequalities by λ and $(1 - \lambda)$ respectively and adding we get $\lambda f(x_1) + (1 - \lambda)f(x_2) \geq f(\lambda x_1 + (1 - \lambda)x_2)$. This is true for each $\lambda \in (0,1)$ and each $x_1, x_2 \in$ ri dom f. Hence f is convex on ri dom f and the proof is complete.

Corollary. Suppose that f is real valued convex function defined on each point in E_n. Then f is convex if and only if f has a subgradient at each $x \in E_n$, i.e., for each $x \in E_n$ there must exist an $\xi \in E_n$ such that $f(y) \geq f(x) + \langle y - x, \xi\rangle$ for all $y \in E_n$.

The reader may note that the above theorem cannot be extended in general to the statement that if for each $x_o \in$ ri dom f there is an ξ with $f(x) \geq f(x_o) + \langle x - x_o, \xi\rangle$ then f is convex on S. An example which shows that the extension is not possible is given below:

$$f(x,y) = \begin{cases} 0 & 0 < y \leq 1, \quad 0 \leq x \leq 1 \\ \frac{1}{4} - (x - \frac{1}{2})^2 & y = 0, \quad 0 \leq x \leq 1 \\ \infty & \text{otherwise} \end{cases}$$

Here dom f = $\{(x,y) : 0 \le x \le 1, 0 \le y \le 1\}$. f has a vector ξ satisfying the inequality $f(x) \ge f(x_o) + \langle x - x_o, \xi\rangle$ for each $x_o \in$ ri dom f (e.g., let $\xi = (0,0)$). However f is not convex on dom f since the epigraph of f is clearly not a convex set.

The following theorem gives the analogue of Theorem 4.1.7 above for the concave case. The proof is similar and is hence omitted.

4.1.8 Theorem. Let S be a convex set in E_n and $f : S \to E_1 \cup \{-\infty\}$. Then f is concave on S only if for each $x_o \in$ ri dom f there exists a vector $\xi \in E_n$ such that $f(x) \le f(x_o) + \langle x - x_o, \xi\rangle$ for every $x \in S$. Conversely, if for each $x_o \in$ ri dom f there is a vector ξ with $f(x) \le f(x_o) + \langle x - x_o, \xi\rangle$ for each $x \in S$ then f is concave of ri dom f.

The above two theorems essentially show that for any point $x_o \in$ ri dom f there is a vector $\xi \in E_n$ such that $(\xi,-1)$ $[(-\xi,1)$ in case of concave function$]$ corresponds to a supporting hyperplane of the epigraph (hypograph) of the function. Here $(\xi,-1)$ $[(-\xi,1)]$ is the normal to the supporting hyperplane which passes through the point $(x_o,f(x_o))$.

4.2. Continuity and Directional Differentiability of Convex Functions

Convex functions has the interesting property of being continuous on the relative interior of their domain as given by Theorem 4.2.1 below.

4.2.1 Theorem. Let $f : S \to E_1$ be a convex function where S is a convex set. Let $x_o \in$ ri S. Then f is continuous at x_o.

Proof: We will show that given any $\epsilon < 0$ we can find a $\delta > 0$ such that $x \in M(S)$ with $\|x - x_o\| < \delta$ implies that $|f(x) - f(x_o)| < \epsilon$. Let L be the linear subspace parallel to M(S) and let $z_1, z_2, \ldots, z_k$ be a basis for L (see Theorem 1.2.4). If $x \in M(S)$ then $x - x_o \in L$ and so $x - x_o = \sum_{i=1}^{k} \alpha_i z_i$. x can be written as $x = x_o + \sum_{i=1}^{k} \alpha_i z_i = \frac{1}{k} \sum_{i=1}^{k} (x_o + k\alpha_i z_i)$ and by convexity of f we get $f(x) \le \frac{1}{k} \sum_{i=1}^{k} f(x_o + k\alpha_i z_i)$. Therefore $f(x) - f(x_o) \le \frac{1}{k} \sum_{i=1}^{k} [f(x_o + k\alpha_i z_i) - f(x_o)]$. Since $x_o \in$ ri S then there is a $\delta' > 0$ such that $x \in M(S)$ with $\|x - x_o\| < \delta'$ implies that $x \in S$.

Consider $x_o + k\alpha_i z_i$ where $\alpha_i \geq 0$ and $\frac{k\alpha_i}{\delta'} \in [0,1]$. By convexity of f we get

$$f(x_o + k\alpha_i z_i) = f\left[\left(1 - \frac{k\alpha_i}{\delta'}\right)x_o + \frac{k\alpha_i}{\delta'}(x_o + \delta' z_i)\right] \leq \left(1 - \frac{k\alpha_i}{\delta'}\right) f(x_o) + \frac{k\alpha_i}{\delta'} f(x_o + \delta' z_i).$$

So $f(x_o + k\alpha_i z_i) - f(x_o) \leq \frac{k\alpha_i}{\delta'}[f(x_o + \delta' z_i) - f(x_o)]$. On the other hand if $\alpha_i < 0$ and $\frac{-k\alpha_i}{\delta'} \in [0,1]$ then $f(x_o + k\alpha_i z_i) \leq \left(1 + \frac{k\alpha_i}{\delta'}\right)f(x_o) - \frac{k\alpha_i}{\delta'} f(x_o - \delta' z_i)$. Therefore $f(x_o + k\alpha_i z_i) - f(x_o) \leq \frac{-k\alpha_i}{\delta'}[f(x_o - \delta' z_i) - f(x_o)]$. Now let $\theta_1 = \max_{1 \leq i \leq k} [f(x_o + \delta' z_i) - f(x_o)]$ and $\theta_2 = \max_{1 \leq i \leq k} [f(x_o - \delta' z_i) - f(x_o)]$. Therefore $f(x_o + k\alpha_i z_i) - (f(x_o) \leq \frac{k|\alpha_i|}{\delta'}\theta$ provided that $\frac{k|\alpha_i|}{\delta'} \in [0,1]$, where $\theta = \max(\theta_1, \theta_2)$. This then means that for $x \in M(S)$ with $x - x_o = \sum_{i=1}^{k} \alpha_i z_i$ we must have $f(x) - f(x_o) \leq \frac{\theta}{\delta'} \sum_{i=1}^{k} |\alpha_i|$, as long as $\frac{k|\alpha_i|}{\delta'} \in [0,1]$. Let $\epsilon > 0$ be given and denote $(z_1, z_2, \ldots, z_k)$ by z. Then $x - x_o = \sum_{i=1}^{k} \alpha_i z_i = z\alpha$ where $\alpha = (\alpha_1, \alpha_2, \ldots, \alpha_k)$. Solving for α we get $\alpha = (z^t z)^{-1} z^t (x - x_o)$, noting that $z^t z$ has an inverse because $z_1, z_2, \ldots, z_k$ are linearly independent. Therefore $\|\alpha\| \leq \|(z^t z)^{-1} z^t\| \, \|x - x_o\| = M \|x - x_o\|$ with an obvious definition of M. This then shows that $|\alpha_i| \leq \|\alpha\| \leq M \|x - x_o\|$ for each i, and $\sum_{i=1}^{k} (\alpha_i) \leq kM\|x - x_o\|$.

Now choose $\delta = \min\left[\delta', \frac{\epsilon\delta'}{kM\theta}, \frac{\delta'}{kM}\right]$. For $x \in M(S)$ with $\|x - x_o\| < \delta$ we have $\frac{k|\alpha_i|}{\delta'} \in [0,1]$ for each i and $f(x) - f(x_o) \leq \frac{\theta}{\delta'} k M\|x - x_o\| \leq \epsilon$. To complete the proof we need to show that $-\epsilon < f(x) - f(x_o)$. Note that $x_o = \frac{1}{2}(x_o + (x - x_o)) + \frac{1}{2}(x_o - (x - x_o))$ and so by convexity of f we have $f(x_o) \leq \frac{1}{2}f(x) + \frac{1}{2}f(2x_o - x)$. Also note that $2x_o - x \in M(S)$ with $\|(2x_o - x) - x_o\| < \delta$ and so we have by the above result $f(2x_o - x) - f(x_o) < \epsilon$. This shows that $f(x_o) \leq \frac{1}{2}f(x) + \frac{1}{2}f(2x_o - x) < \frac{1}{2}f(x) + \frac{1}{2}f(x_o) + \frac{1}{2}\epsilon$, i.e., $-\epsilon < f(x) - f(x_o)$. We have therefore established the fact that given any $\epsilon > 0$ we can find a $\delta > 0$ such that $x \in M(S)$ with $\|x - x_o\| < \delta$ implies that $|f(x) - f(x_o)| < \epsilon$ and the proof is complete.

<u>Corollary 1</u>. Let f be a finite convex function defined everywhere in E_n. Then

f is continuous.

Corollary 2. Let $f : S \to E_1 \cup \{\infty\}$ where S is convex. Then f is continuous on ri dom f.

From the above theorem we know that a convex function is continuous in the relative interior of its effective domain. However, it is clear from Figure 4.1 that a convex function need not be continuous on its effective domain. Points where problems may arise are actually points of the relative boundary of the effective domain of the function. To explore this notion further we consider the following definition of the lower and upper limit functions.

4.2.2 Definition. Let $f : S \to E_1 \cup \{\infty\}$. The lower closure of f, denoted by $\underline{f}$, is defined on $c\ell\, S$ with $\underline{f}(x) = \underline{\lim}_{z \to x} f(z)$ for each $x \in c\ell\, S$. Similarly let $g : S \to E_1 \cup \{-\infty\}$. The upper closure of g, denoted by $\bar{g}$, is defined on $c\ell\, S$ with $\bar{g}(x) = \overline{\lim}_{z \to x} g(z)$ for each $x \in c\ell\, S$.

From the above definition the following facts follow easily:

1. Lower semi-continuity of f at $x \in S$ reduces to $f(x) = \underline{f}(x)$.
2. Upper semi-continuity of g at $x \in S$ reduces to $g(x) = \bar{g}(x)$.
3. Continuity of f at $x \in S$ is equivalent to $f(x) = \underline{f}(x) = \bar{f}(x)$.
4. Let $\{x_n\}$ be a sequence in S which converges to $x \in S$. Then

$$\underline{\lim}_{k \to \infty} f(x_k) = \lim_{K \to \infty} \inf \{f(x_k) : k > K\} \geq \underline{f}(x) \text{ and } \overline{\lim}_{k \to \infty} f(x_k) =$$

$$\lim_{K \to \infty} \sup \{f(x_k) : k > K\} \leq \bar{f}(x).$$

5. Let $x \in \operatorname{dom} \underline{f}$. Then there exists a sequence $\{x_k\}$ converging to x such that $f(x_k) \to \underline{f}(x)$.
6. Let $x \in \operatorname{dom} \bar{f}$. Then there exists a sequence $\{x_k\}$ converging to x such that $f(x_k) \to \bar{f}(x)$.

Actually one can find convex functions (concave functions) which are neither lower nor upper semi-continuous at a given point. As an example let $f(x) = x^2$ for $|x| < 1$, $f(1) = f(-1) = 2$, and $f(x) = \infty$ for $|x| > 1$. Clearly $\underline{f}(1) = \underline{f}(-1) = 1$ and $\bar{f}(1) = \bar{f}(-1) = \infty$. On the other hand $f(1) = f(-1) = 2$.

In many instances one would like to replace the function f by the lower semi-continuous function $\underline{f}$, which is better behaving at points of the relative boundary. Similarly, for a concave function g, one would like to replace it by the upper semi-continuous function $\bar{g}$.

The reader may get the impression that a closed convex function is continuous on its domain. Even though this is true for functions of one variable, it is not true in general. This fact is indicated by the following example:

$$f(x,y) = \begin{cases} 0 \text{ for } x = y = 0 \\ \frac{y^2}{x} \text{ for } x > 0 \\ \infty \text{ otherwise} \end{cases}$$

Convexity of f can be easily verified from the identity $(\lambda y_1 + (1 - \lambda)y_2)^2 = (\lambda x_1 + (1 - \lambda)x_2)\left(\frac{\lambda y_1^2}{x_1} + \frac{(1 - \lambda)y_2^2}{x_2}\right) - \lambda(1 - \lambda)\frac{(x_1y_2 - x_2y_1)^2}{x_1x_2}$. Clearly f is not continuous on dom f since f is not continuous at $(x,y) = (0,0)$. Noting that $\underline{f} = f$ it is then clear that $\underline{f}$ is not continuous of dom f.

The following theorem gives an initially clear fact, namely the epigraph of the lower closure of a function is indeed the closure of the epigraph of the function. Here no convexity is needed.

4.2.3 Theorem. $E_{\underline{f}} = c\ell\, E_f$ and $H_{\bar{g}} = c\ell\, H_g$.

Proof: We will first show that $E_{\underline{f}} \subset C\ell\, E_f$. Let $(x,y) \in E_{\underline{f}}$, i.e., $y \geq \underline{f}(x)$ and $x \in C\ell\, S$. This by definition means that $y \geq \underline{f}(x) = \liminf_{z \to x} f(z)$, i.e., there exists a sequence $\{z_k\}$ is S converging to x such that $\underline{f}(x) = \lim_{k \to \infty} f(z_k)$. This implies that there exists a sequence of scalors $\{\epsilon_k\}$ converging to zero such that $\underline{f}(x) + \epsilon_k = f(z_k)$ for all k. This in turn implies that $y + \epsilon_k \geq f(z_k)$ for all k, and $\epsilon_k \to 0$, $z_k \to x$. Letting $y_k = y + \epsilon_k$, then we have constructed a sequence $(z_k, y_k) \in E_f$ with limit (x,y) which implies that $(x,y) \in C\ell\, E_f$. This shows that $E_{\underline{f}} \subset C\ell\, E_f$. To show the converse inclusion let $(x,y) \in C\ell\, E_f$, i.e., $(x,y) = \lim(x_k, y_k)$ with $y_k \geq f(x_k)$, $x_k \in S$. This shows that $y = \lim y_k \geq \limsup_{x_k \to x} f(x_k) \geq \liminf_{x_k \to x} f(x_k) \geq \liminf_{z \to x} f(z)$ $= \underline{f}(x)$, i.e., $(x,y) \in E_{\underline{f}}$. This shows that $E_{\underline{f}} = C\ell\, E_f$. The case of the hypograph of the upper limit can be similarly proved.

Corollary. Let f be a convex function defined on a convex set S. Then $\underline{f}$ is a convex function on $C\ell\, S$. Similarly, if g is concave on S then $\bar{g}$ is concave on $C\ell S$.

We will now discuss the notion of directional differentiability of a convex function. We will first present the definitions of directional differentiability and one-sided directional differentiability of an arbitrary function. We later show that a convex function has a one-sided directional derivative everywhere in its effective domain.

4.2.4 Definition. Let $f : S \to E_1 \cup \{\infty\}$ and let $x_o \in S$ be such that $f(x_o) < \infty$. Let x be any vector in E_n with $\|x\| = 1$. If 0 is an interior point of the set of real t for which $x_o + t\,x \in S$ and the limit, $\lim_{t\to 0} \frac{f(x_o + t\,x) - f(x_o)}{t}$ exists, we say that f has a directional derivative at x_o in the direction x, and the value of this limit, which is denoted by $f'(x_o;x)$ is called the directional derivative of f in the direction x. f is said to have a one-sided directional derivative at x_o in the direction x if $\lim_{t\to 0^+} \frac{f(x_o + t\,x) - f(x_o)}{t}$ exists. The one-sided directional derivative is denoted by $f'_+(x_o;x)$.

Needless to say that the requirement $\|x\| = 1$ can be deleted as long as x is not the zero vector. Now in the case of a convex function which is defined on a subset S of E_n we can extend the function to E_n by letting $f(x) = \infty$ for $x \notin S$. This is convenient because we do not have to worry then whether 0 is an interior point of real t satisfying $x_o + t\,x \in S$ or not. The following theorem shows that for a convex function the one-sided directional derivative indeed exists at every point in the effective domain at any direction. Here we certainly allow $+\infty$ and $-\infty$ as limits.

4.2.5 Theorem. Let $f : E_n \to E_1 \cup \{\infty\}$ be convex and let $x_o \in \text{dom } f$. Then $f'_+(x_o;x)$ exists for each nonzero $x \in E_n$.

Proof: We first show that $\frac{f(x_o + \lambda x) - f(x_o)}{\lambda}$ is a nondecreasing function of $\lambda > 0$. Let $\lambda_2 > \lambda_1 > 0$ and write $x_o + \lambda_1 x = \frac{\lambda_1}{\lambda_2}(x_o + \lambda_2 x) + \left(1 - \frac{\lambda_1}{\lambda_2}\right) x_o$ where $\frac{\lambda_1}{\lambda_2} \in (0,1)$. By convexity of f we must have $f(x_o + \lambda_1 x) \le \frac{\lambda_1}{\lambda_2} f(x_o + \lambda_2 x) + \left(1 - \frac{\lambda_1}{\lambda_2}\right) f(x_o)$. This shows that $\frac{f(x_o + \lambda_1 x) - f(x_o)}{\lambda_1} \le \frac{f(x_o + \lambda_2 x) - f(x_o)}{\lambda_2}$. This shows that $\lim_{\lambda\to 0^+} \frac{f(x_o + \lambda x) - f(x_o)}{\lambda} = \inf_{\lambda>0} \frac{f(x_o + \lambda x) - f(x_o)}{\lambda}$ indeed exists and the proof is complete.

4.3. Differentiable Convex Functions

We will now turn our attention to differentiable convex functions. Since differentiability of a function at a point x_c requires the point x_o to be in the interior of the set under consideration, we will assume that the set under consideration is open. We will also consider convex real valued functions. Noting that a convex real valued function f defined on a set S will yield a convex function if we extend it by letting $f(x) = \infty$ for $x \notin S$, one can easily consider differentiable convex functions which may assume infinite values.

Lemma 4.3.1 below shows that in the case of a differentiable convex function f at x_o, any subgradient ξ (see Theorem 4.1.7) is indeed the gradient vector $\nabla f(x_o)$. In other words the subgradient is unique, namely the gradient vector.

4.3.1 Lemma. Let $f : S \to E_1$ be convex and let S be an open convex set in E_n. Let ξ be a subgradient of f at $x_o \in S$, i.e., $f(x) \geq f(x_o) + \langle x - x_o, \xi\rangle$ for each $x \in S$. Then $\xi = \nabla f(x_o)$ provided that the latter exists.

Proof: Consider the sequence $x_k = x_o + \frac{1}{k}(\xi - \nabla f(x_o))$. Since $x_o \in S$ and S is open then $x_k \in S$ for k sufficiently large. By differentiability of f at x_o we have $f(x_k) = f(x_o) + \langle x_k - x_o, \nabla f(x_o)\rangle + \|x_k - x_o\| . \epsilon(x_k - x_o)$ where $\epsilon(t) \to 0$ as $t \to 0$. Since ξ is a subgradient of f at x_o we also have $-f(x_k) \leq -f(x_o) - \langle x_k - x_o, \xi\rangle$. Adding the last two expressions we get

$$0 \leq \frac{1}{k} \langle x_k - x_o, \nabla f(x_o) - \xi\rangle + \|x_k - x_o\| \cdot \epsilon(x_k - x_o)$$

$$= \frac{1}{k} \langle \xi - \nabla f(x_o), \nabla f(x_o) - \xi\rangle + \frac{1}{k}\|\xi - \nabla f(x_o)\| \cdot \epsilon\left(\frac{1}{k}(\xi - \nabla f(x_o)\right) \quad \text{for}$$

k sufficiently large. Multiplying by k and letting $k \to \infty$ we get $0 \leq - \|\nabla f(x_o) - \xi\|^2$, which implies that $\nabla f(x_o) = \xi$. This completes the proof.

The reader may note the following implications of the above remark. For a convex function f with a subgradient ξ at a point x_o there corresponds a supporting hyperplane to the epigraph of f at the point $(x_o, f(x_o))$. Moreover, the hyperplane has a normal vector $(\xi, -1)$. Since for a differentiable function we have a unique subgradient at each point, namely the gradient vector, then a differentiable function has a unique supporting hyperplane at each boundary point of its epigraph.

Now suppose that $f : S \to E_1$ be convex and differentiable at $x_o \in S$. Here S is an open convex set. Then by Theorem 4.1.7 we conclude that f has at least one subgradient ξ at x_o. By means of the above remark it is then clear that $\xi = \nabla f(x_o)$, i.e., $f(x) \geq f(x_o) + \langle x - x_o, \nabla f(x_o)\rangle$ for each $x \in S$. Thus the following theorem is obvious.

4.3.2 Theorem. Let $f : S \to E_1$ be differentiable on S where S is an open convex set. Then f is convex on S if and only if $f(x) \geq f(x_o) + \langle x - x_o, \nabla f(x_o)\rangle$ for each $x \in S$ for any arbitrary $x_o \in S$.

The following theorem gives another important characterization of convex functions.

4.3.3 Theorem. Let $f : S \to E_1$ be differentiable on S where S is an open convex set. Then f is convex on S if and only if $\langle x_2 - x_1, \nabla f(x_2) - \nabla f(x_1)\rangle \geq 0$ for each $x_1, x_2 \in S$.

Proof: Assume that f is convex on S and let $x_1, x_2 \in S$. Thus by Theorem 4.3.2 above we must have $f(x_2) \geq f(x_1) + \langle x_2 - x_1, \nabla f(x_1)\rangle$ and $f(x_1) \geq f(x_2) + \langle x_1 - x_2, \nabla f(x_2)\rangle$. Adding these two inequalities we get $\langle x_2 - x_1, \nabla f(x_1) - \nabla f(x_2)\rangle \leq 0$. To prove the converse let x_1, x_2 be arbitrary points of S. Then by the mean value theorem, $f(x_2) - f(x_1) = \langle x_2 - x_1, \nabla f(x_\lambda)\rangle$ for some $x_\lambda = (1 - \lambda)x_1 + \lambda x_2$ where $\lambda \in (0,1)$. But by the hypothesis of the theorem, $\langle \nabla f(x_\lambda) - \nabla f(x_1), x_\lambda - x_1\rangle \geq 0$, i.e., $\lambda\langle \nabla f(x_\lambda) - \nabla f(x_1), x_2 - x_1\rangle \geq 0$. This implies that $\langle \nabla f(x_\lambda), x_2 - x_1\rangle \geq \langle \nabla f(x_1), x_2 - x_1\rangle$ Hence $f(x_2) \geq f(x_1) + \langle x_2 - x_1, \nabla f(x_1)\rangle$ and f is convex in view of Theorem 4.3.2 above. This completes the proof.

So far we have considered once differentiable functions. Twice differentiable convex functions have an important characterization. This characterization is helpful for checking convexity of numerous twice differentiable functions.

Let f be a real valued function defined on a set S. S is said to be twice differentiable at a point $x_o \in \text{int } S$ if $f(x) = f(x_o) + \langle x - x_o, \nabla f(x_o)\rangle + \frac{1}{2}(x - x_o)^t M(x - x_o) + \|x - x_o\|^2 \cdot \epsilon(x - x_o)$ where $\nabla f(x_o)$ is the gradient of f at x_o, M is an nxn matrix (M is actually the Hessian matrix of f at x_o where the ij^{th} entry is $\frac{\partial^2 f}{\partial x_i \partial x_j}$ evaluated at x_o), and $\epsilon(t) \to 0$ as $t \to 0$.

4.3.4 Theorem. Let $f : S \to E_1$ be twice differentiable on S, where S is an open convex set. Then f is convex on S if and only if the Hessian matrix of f is positive semi-definite everywhere on S.

Proof: Suppose that f is convex on S and let $x_o \in S$. Let $x \in E_n$. We want to show that $x^t Mx \geq 0$ where M is the Hessian matrix of f at x_o. Since S is open and $x_o \in S$ then $x_o + \lambda x \in S$ for $\lambda \in [-\delta, \delta]$ where δ is some positive number. By convexity of f (see Theorem 4.3.2) $f(x_o + \lambda x) \geq f(x_o) + \lambda\langle x, \nabla f(x_o)\rangle$. Also by twice differentiability of f at x_o we must have $f(x_o + \lambda x) = f(x_o) + \lambda\langle x, \nabla f(x_o)\rangle + \frac{1}{2}\lambda^2 x^t Mx + \lambda^2 \|x\|^2 \cdot \epsilon(\lambda x)$. From the last two expressions we get $\frac{1}{2}\lambda^2 x^t Mx + \lambda^2 \|x\|^2 \cdot \epsilon(\lambda x) \geq 0$. Dividing by $\lambda^2 > 0$ and letting $\lambda \to 0$, $x^t Mx \geq 0$. To show the converse suppose that the Hessian matrix of f is positive semi-definite everywhere on S. Now let $x_1, x_2 \in S$. Then by Taylor's theorem $f(x_2) = f(x_1) + \langle x_2 - x_1, \nabla f(x_1)\rangle + \frac{1}{2}(x_2 - x_1)^t M_\lambda (x_2 - x_1)$ where M_λ is the Hessian matrix of f evaluated at some point $\lambda x_1 + (1 - \lambda)x_2$ where $\lambda \in (0,1)$. By assumption M_λ is positive semi-definite and so $(x_2 - x_1)^t M_\lambda (x_2 - x_1) \geq 0$. This shows that $f(x_2) \geq f(x_1) + \langle x_2 - x_1, \nabla f(x_1)\rangle$, and convexity of f on S follows from Theorem 4.3.2. This completes the proof.

As an example let us consider the function f defined by $f(x,y) = x^2 + y^2$ if $0 < x < 1$, $0 < y < 1$, and ∞ otherwise. We would like to check convexity of f by means of the above theorem. Let us consider the restriction of f to the open set $\{(x,y) : 0 < x < 1, 0 < y < 1\}$. This function is convex in the described set if and only if the Hessian matrix is positive semi-definite everywhere. This is indeed the case since the matrix $\begin{bmatrix} 2 & 0 \\ 0 & 2 \end{bmatrix}$ is positive definite and so convexity of the restriction of f to $\{(x,y) : 0 < x < 1, 0 < y < 1\}$ follows. Since the epigraph of this function and the epigraph of f are equal then convexity of f is immediate.

In many cases the function under consideration is not finite on an open set and hence direct application of Theorem 4.3.4 above may not be possible. For example, consider a slight modification of the above problem where $f(x,y) = x^2 + y^2$ if $0 \leq x \leq 1$, $0 \leq y \leq 1$, and ∞ otherwise. We want to check convexity of this function. This can be approached as follows. Consider the function $g(x,y) = x^2 + y^2$ for $0 < x < 1$ and $0 < y < 1$.

Convexity of g was verified by Theorem 4.3.4 above. By definition of the closure of a convex function it is clear that $\underline{g}(x,y) = x^2 + y^2$ for $0 \le x \le 1$ and $0 \le y \le 1$. By the corollary in Theorem 4.2.3 it is clear that $\underline{g}$ is convex [Note that E_g is a convex set and so is $C\ell\, E_g$ which is $E_{\underline{g}}$ by Theorem 4.2.3. But since the epigraph of $\underline{g}$ is a convex set then $\underline{g}$ is a convex function on $\{(x,y) : 0 \le x \le 1, 0 \le y \le 1\}$ by Lemma 4.1.3]. But since the epigraphs of $\underline{g}$ and the function f are identical then the function f is convex (since E_f is convex).

4.4. Some Examples of Convex Functions

In this section we will discuss some important convex functions which arise in different contexts. We will discuss the indicator function, the distance function, the gauge function, and the support function.

Let us start with the distance function. The reader may recall that given an arbitrary set S in E_n then we define the distance function $d(\cdot,S)$ by $d(x,S) = \inf\{\|x - y\| : y \in S\}$. We will now show that given a convex set S then $d(\cdot,S)$ is indeed a convex function on E_n. The reader may recall that given any point $x \notin C\ell\, S$ there is a unique minimizing point (see Theorem 2.4.4). In other words $d(x,S) = d(x,C\ell S) = \min\{\|x-y\| : y \in C\ell\, S\} = \|x - x_0\|$ where $x_0 \in C\ell\, S$ is unique. On the other hand if $x \in C\ell\, S$ then clearly $d(x,S) = 0$ and x_0 is x itself. We will make use of this information in showing that the distance function is convex. Then we need to show that $d(\lambda x + (1 - \lambda)y,S) \le \lambda\, d(x,S) + (1 - \lambda)\, d(y,S)$ for each $\lambda \in (0,1)$. But $d(\lambda x + (1 - \lambda)y,S) = d(\lambda x + (1 - \lambda)y,C\ell\, S) \le \|\lambda x + (1 - \lambda)y - \lambda\hat{x} - (1 - \lambda)\hat{y}\|$, where $d(x,S) = \|x - \hat{x}\|$ and $d(y,S) = \|y - \hat{y}\|$. But by convexity of S, $C\ell\, S$ is convex, and so $\lambda\hat{x} + (1 - \lambda)\hat{y} \in C\ell\, S$ for each $\lambda \in (0,1)$. Therefore $d(\lambda x + (1 - \lambda)y,S) \le \|\lambda(x - \hat{x}) + (1 - \lambda)(y - \hat{y})\| \le \lambda\|x - \hat{x}\| + (1 - \lambda)\|y - \hat{y}\| = \lambda\, d(x,S) + (1 - \lambda)\, d(y,S)$, and convexity of $d(\cdot,S)$ follows.

We will consider two numerical examples here. First let $S = \{x : -1 \le x \le 1\}$. Clearly $d(x,S) = 0$ for $-1 \le x \le 1$, $d(x,S) = x - 1$ for $x > 1$, and $d(x,S) = -1 - x$ for $x < -1$. The distance function is illustrated in Figure 4.2 and $d(\cdot,S)$ is clearly a convex function.

As another example let $S = \{(x,y) : x^2 + y^2 \le 1\}$. This set is given in Figure 4.3 below. It is obvious that $d(x,y),S) = 0$ as long as $x^2 + y^2 \le 1$. On the other

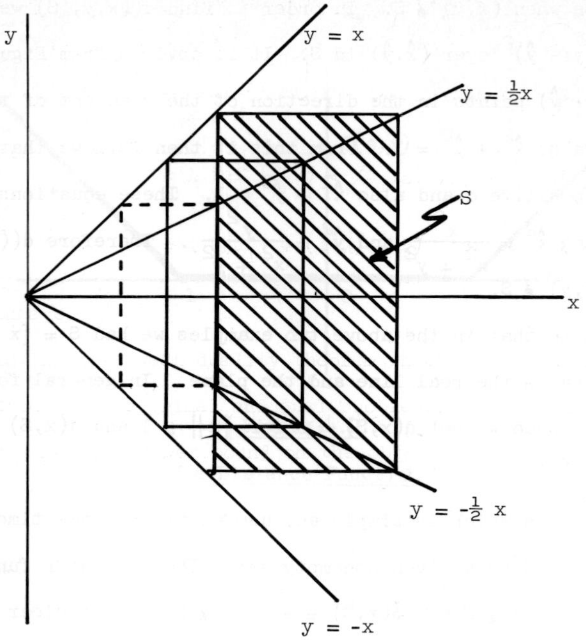

Figure 4.4

An Example of Gauge Function

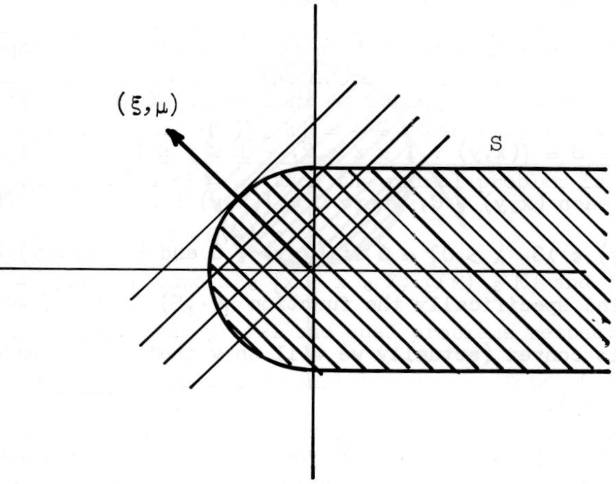

Figure 4.5

An Example of the Support Function

or $\gamma(x_2,S) = \infty$. So suppose that $\gamma(x_1,S)$ and $\gamma(x_2,S)$ are finite. Given $\epsilon > 0$, where λ_1 and λ_2 such that $\gamma(x_1,S) < \lambda_1 < \gamma(x_1,S) + \epsilon$ and $\gamma(x_2,S) < \lambda_2 < \gamma(x_2,S) + \epsilon$. Note that $\lambda_1 > \gamma(x_1,S) = \inf\{\lambda : x_1 \in \lambda S, \lambda > 0\}$, i.e., $x_1 \in \lambda_1 S$ or $\frac{x_1}{\lambda_1} \in S$. Similarly $\frac{x_2}{\lambda_2} \in S$. By convexity of S we must have $\frac{\lambda_1}{\lambda_1 + \lambda_2}\frac{x_1}{\lambda_1} + \frac{\lambda_2}{\lambda_1 + \lambda_2}\frac{x_2}{\lambda_2} \in S$, i.e.,

$\frac{1}{\lambda_1 + \lambda_2}(x_1 + x_2) \in S$. Therefore $\gamma(x_1 + x_2,S) = \inf\{\lambda : x_1 + x_2 \in \lambda S\} \le \lambda_1 + \lambda_2$ since $x_1 + x_2 \in (\lambda_1 + \lambda_2)S$. But by our choice $\lambda_1 + \lambda_2 < \gamma(x_1,S) + \gamma(x_2,S) + 2\epsilon$. This means that $\gamma(x_1 + x_2,S) \le \gamma(x_1,S) + \gamma(x_2,S) + 2\epsilon$. The result follows since the last inequality holds for each $\epsilon > 0$.

Needless to say that positive homogeneity and subadditivity imply convexity. This is verified immediately since $\gamma(\lambda x_1 + (1 - \lambda)x_2,S) \le \gamma(\lambda x_1,S) + \lambda(1 - \lambda)x_2,S)$ $= \lambda\gamma(x_1,S) + (1 - \lambda)\gamma(x_2,S)$ for each $\lambda \in (0,1)$.

Let us turn to our example discussed before. We found out that $\gamma((x,y),S) = \infty$ for $(x,y) \notin C$. We can use positive homogeneity in finding γ on C once we find γ on the line segment $\{(x,y) : x = 1\}$. The reader may easily check that for $x = 1$,

$$\gamma((x,y),S) = \begin{cases} \frac{1}{2} & \text{if } |y| \le \frac{1}{2} \\ |y| & \text{if } \frac{1}{2} \le |y| \le 1 \\ \infty & \text{if } |y| > 1 \end{cases}$$

Now we can find $\gamma((x,y),S)$ on C. $\gamma((x,y),S) = \gamma\left(x\left(1,\frac{y}{x}\right),S\right)$

$$= x\gamma\left(\left(1,\frac{y}{x}\right),S\right) = \begin{cases} \frac{1}{2}x & \text{if } \left|\frac{y}{x}\right| \le \frac{1}{2} \text{ and } x > 0 \\ x|y| & \text{if } \frac{1}{2} \le \left|\frac{y}{x}\right| \le 1 \text{ and } x > 0 \\ \infty & \text{if } \left|\frac{y}{x}\right| > 1 \text{ and } x > 0 \\ \infty & \text{if } x \le 0 . \end{cases}$$

For $x = 0$, it is obvious that $\gamma((0,0),S) = 0$. This completely defines the function γ on E_n.

Now we would like to discuss <u>support functions</u> of convex sets. Let S be a nonempty convex set in E_n. The support function $\delta^*(\cdot,S)$ is defined on E_n by $\delta^*(\xi,S) = \sup_{x \in S} \langle \xi,x\rangle$. Convexity of the support function can be easily checked. Actually $\delta^*(\cdot,S)$ is positively homogeneous and subadditive. The support function corresponds to the notion of supporting hyperplanes of convex sets at boundary points. To illustrate this point let us consider the example shown in Figure 4.5. Here $S = \{(x,y) : x < 0, x^2 + y^2 \leq 1\} \cup \{(x,y) : x \geq 0, -1 \leq y \leq 1\}$. $\delta^*((\xi,\mu),S) = \sup_{(x,y) \in S} \xi x + \mu y$. Clearly if $\xi > 0$ then the sup is equal to infinity since x can be chosen arbitrarily large. On the other hand if $\xi < 0$ then the sup will obviously be achieved at the boundary of the half disk. Actually we like to maximize $\xi\sqrt{1-y^2} + \mu y$. Using elementary algebra we conclude that $(x,y) = \left(\frac{\xi}{\sqrt{\xi^2 + \mu^2}}, \frac{\mu}{\sqrt{\xi^2 + \mu^2}}\right)$ and $\delta^*((\xi,\mu),S) = \sqrt{\xi^2 + \mu^2}$. As shown in Figure 4.5 the case where $\xi \leq 0$ corresponds to a supporting hyperplane of the set S having normal vector (ξ,μ). When $\xi > 0$ there is no supporting hyperplane of S with normal vector (ξ,μ) and so $\delta^*((\xi,\mu),S) = \infty$.

It is worthwhile mentioning that the indicator function and the support function have an important conjugacy relationship in common. Actually it is clear by the definition of both functions that $\delta^*(x,S) = \sup_{y \in E} \{\langle y,x\rangle - \delta(y,S)\}$. Since $\delta(y,S) = \infty$ for $y \notin S$ then the sup above can be restricted to S which means that $\delta^*(x,S) = \sup_{y \in S} \langle y,x\rangle$. The notion of conjugacy and conjugate functions will be discussed later in Chapter 9.

As an example of a convex function we consider the supremum of a collection of convex functions. These functions arise in many contexts, e.g., min max problems. Let $\{f_i\}_{i \in I}$ be a collection of convex functions defined on a convex set S. Let $f = \sup_{i \in I} f_i$, i.e., $f(x) = \sup_{i \in I} f_i(x)$ for each $i \in I$. We show that f is convex. Let $x,y \in S$ then $f(\lambda x + (1 - \lambda)y) = \sup_{i \in I} f_i(\lambda x + (1 - \lambda)y) \leq \sup_{i \in I}[\lambda f_i(x) + (1 - \lambda)f_i(y)$ $\leq \lambda \sup_{i \in I} f_i(x) + (1 - \lambda) \sup_{i \in I} f_i(y) = \lambda f(x) + (1 - \lambda)f(y)$ for each $\lambda \in (0,1)$ so f is convex.

Now consider the case when f is the maximum of two differentiable convex functions f_1 and f_2 which are defined on an open convex set S. We will have different occasions to deal with such a function. Here we are interested to find the form of subgradients of this new function f. So let $x_o \in S$ be such that $f(x_o) = f_1(x_o) = f_2(x_o)$. It can be immediately checked that $E_f = E_{f_1} \cap E_{f_2}$. Let $E_f' = E_f - (x_o,f(x_o))$, $E_{f_1}' = E_{f_1} - (x_o,f(x_o))$, and $E_{f_2}' = E_{f_2} - (x_o,f(x_o))$. Therefore $E_f' = E_{f_1}' \cap E_{f_2}'$. Note that E_{f_1}' and E_{f_2}' have nonempty interiors, e.g., $(0,\delta) \in \text{int } E_{f_2}' \cap \text{int } E_{f_2}'$ for any $\delta > 0$. Therefore by Theorem 3.1.14 we must have $(E_f')^* = (E_{f_1}' \cap E_{f_2}')^* = (E_{f_1}')^* +$

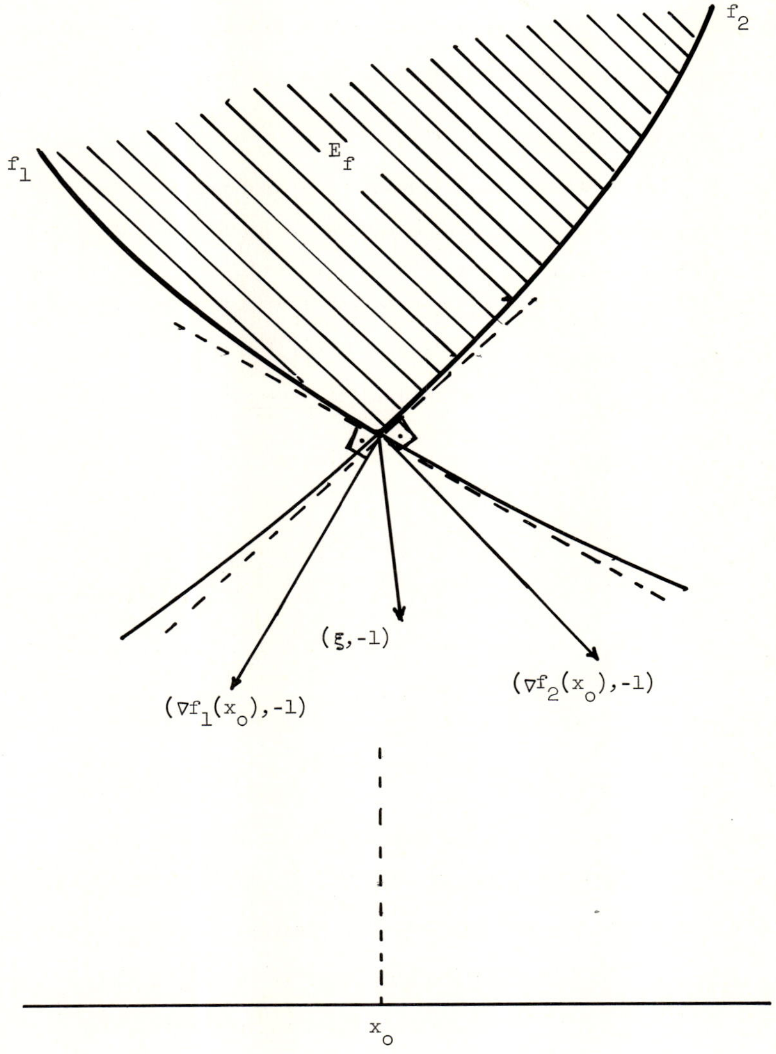

Figure 4.6

An Example of the Sup Function

$(E'_{f_2})^*$. However $(E'_f)^* = \{\lambda(\xi,-1) : \lambda \geq 0,\ f(x) \geq f(x_0) + \langle x - x_0, \xi\rangle\}$ whereas $E'_{f_1} = \{\lambda\nabla f_1(x_0), -1) : \lambda \geq 0\}$ and $E'_{f_2} = \{\lambda(\nabla f_2(x_0), -1) : \lambda \geq 0\}$. This follows by definition of a subgradient of a convex function and by the fact that any supporting hyperplane to the epigraph of a convex function at a point corresponding to the interior of the domain is nonvertical. This then establishes the fact that a vector ξ is a subgradient of f at x_0 if and only if $\xi = \lambda\nabla f_1(x_0) + (1 - \lambda)\nabla f_2(x_0)$ where $\lambda \in (0,1)$. The situation is illustrated in Figure 4.6.

4.5. Generalization of Convex Functions

In this section we will discuss quasi-convex and pseudo-convex functions. We will also briefly discuss the important notion of convexity at a point.

4.5.1 Definition. Let $f : S \to E_1 \cup \{\infty\}$ where S is a convex set in E_n. f is said to be quasi-convex if for each $x_1, x_2 \in S$ with $f(x_1) \leq f(x_2)$ we have $f(\lambda x_1 + (1 - \lambda)x_2 \leq f(x_2)$ for every $\lambda \in (0,1)$. f is said to be strictly quasi-convex if for each $x_1, x_2 \in S$ with $f(x_1) < f(x_2)$ we have $f(\lambda x_1 + (1 - \lambda)x_2) < f(x_2)$ for each $\lambda \in (0,1)$.

A function f defined on a convex set S with values in $E_1 \cup \{-\infty\}$ is said to be quasi-concave if - f is quasi-convex. In other words, $x_1, x_2 \in S$ with $f(x_2) \geq f(x_1)$ implies that $f(\lambda x_1 + (1 - \lambda)x_2) \geq f(x_1)$ for each $\lambda \in (0,1)$. Strict quasi-concavity is defined in a similar fashion.

An example of a quasi-convex function is $f(x) = x^3$ for $x \in E_1$. One can easily establish a characterization of quasi-convex functions via their level sets. This is given by Theorem 4.5.2 below from which quasi-convexity of x^3 follows.

4.5.2 Theorem. Let $f : S \to E_1 \cup \{\infty\}$ where S is a convex set in E_n. f is quasi-convex on S if and only if $S_\alpha = \{x \in S : f(x) \leq \alpha\}$ is convex for each real number α.

Proof: Suppose f is quasi-convex on S. Consider S_α for some arbitrary real number α. Let $x_1, x_2 \in S_\alpha$. We want to show that $\lambda x_1 + (1 - \lambda)x_2 \in S_\alpha$ for every $\lambda \in (0,1)$. Let $\beta = \max\,(f(x_1), f(x_2)) \leq \alpha$. By quasi-convexity of f on S, then $f(\lambda x_1 + (1 - \lambda)x_2) \leq \beta \leq \alpha$. Hence $\lambda x_1 + (1 - \lambda)x_2 \in S_\alpha$ for each $\lambda \in (0,1)$. Conversely, let S_α be convex for each α and let $x_1, x_2 \in S$ with $f(x_1) \leq f(x_2)$. Then $x_1, x_2 \in S_{f(x_2)}$ and hence by convexity of $S_{f(x_2)}$ it follows that $\lambda x_1 + (1 - \lambda)x_2 \in S_{f(x_2)}$ for each $\lambda \in (0,1)$, i.e., $f(\lambda x_1 + (1 - \lambda)x_2) \leq f(x_2)$. This completes the proof.

As seen in Section 4.1, strict convexity implies convexity, i.e., every strictly convex function is indeed a convex function. However, in general, strict quasi-convexity does not imply quasi-convexity. Needless to say that quasi-convexity does not imply strict quasi-convexity either. For example, consider f defined by $f(x) = \begin{cases} 1 & x = 0 \\ 0 & x \neq 0 \end{cases}$. f is strictly quasi-convex on E_1 but not quasi-convex on E_1. Now consider f defined by $f(x) = \begin{cases} |x|, & x < 1 \\ 1, & x \geq 1 \end{cases}$. f is quasi-convex on E_1 but not strictly quasi-convex on E_1.

From the above, the name strict quasi-convexity may sound erroneous in the classical use of the word strict. However, the name is justified when f is lower semi-continuous. In this case, strict quasi-convexity implies quasi-convexity. This fact is proved below.

4.5.3 Theorem. Let S be a nonempty convex in E_n. Let $f : S \to E_1 \cup \{\infty\}$ be lower semi-continuous and strictly quasi-convex on S. Then f is quasi-convex on S.

Proof: Let $x_1, x_2 \in S$. If $f(x_2) < f(x_1)$ then by strict quasi-convexity of f we get $f(\lambda x_1 + (1 - \lambda)x_2) < f(x_1)$ for each $\lambda \in (0,1)$. Now suppose $f(x_2) = f(x_1)$. Consider the set $L = \{x : x = \lambda x_1 + (1 - \lambda)x_2, \lambda \in (0,1), f(x) > f(x_1)\}$. f is quasi-convex if and only if $L = \emptyset$ for each $x_1, x_2 \in S$. Suppose on the contrary that $L \neq \emptyset$. Let $\hat{x} \in L$, i.e., $f(\hat{x}) > f(x_1)$, $\hat{x} = \lambda x_1 + (1 - \lambda)x_2$ for some $\lambda \in (0,1)$. Since f is lower semi-continuous on S, then there exists an $x_0 \in L$ such that $x_0 = \mu x_1 + (1 - \mu)\hat{x}$ for some $\mu \in (0,1)$. By strict quasi-convexity of f, we conclude that $f(x_1) < f(\hat{x})$ implies $f(x_0) < f(\hat{x})$, and $f(x_2) = f(x_1) < f(x_0)$ implies $f(\hat{x}) < f(x_0)$, a contradiction, then $L = \emptyset$ and f is quasi-convex on S. This completes the proof.

The following theorem shows the significance of strict quasi-convexity in nonlinear programming. This property insures that a local minimum is also a global minimum.

4.5.4 Theorem. Let S is a nonempty convex set in E_n and let $f : S \to E_1$ be strictly quasi-convex. Let N be an open ball about $x_0 \in S$ such that $f(x_0) \leq f(x)$ for all $x \in S \cap N$ (i.e., x_0 is a local minimum of f over S). Then $f(x_0) \leq f(x)$ for each $x \in S$.

Proof: Assume, on the contrary, that there exists a $\hat{x} \in S$ such that $f(\hat{x}) < f(x)$. By convexity of S, there exists a $\delta \in (0,1)$ such that $\lambda x_0 + (1 - \lambda)\hat{x} \in S \cap N$ for each $\lambda \in (0,\delta)$. Hence by hypothesis $f(x_o) \leq f(\lambda x_o + (1 - \lambda)\hat{x})$ for each $\lambda \in (0,\delta)$. But this contradicts the fact that f is strictly quasi-convex since then $f(\hat{x}) < f(x_o)$ implies $f(\lambda x_o + (1 - \lambda)\hat{x}) < f(x_o)$ for each $\lambda \in (0,1)$. This completes the proof.

We will now turn our attention to differentiable quasi-convex functions. We will give a necessary and sufficient characterization of such functions via the gradient vector.

4.5.5 Theorem. Let S be an open convex set and let $f : S \to E_1$ be differentiable on S. f is quasi-convex on S if and only if $x_1, x_2 \in S$ with $f(x_1) \leq f(x_2)$ implies that $\langle x_1 - x_2, \nabla f(x_2)\rangle \leq 0$.

Proof: Let f be quasi-convex on S and let $x_1, x_2 \in S$ be such that $f(x_1) \leq f(x_2)$. Therefore $f(\lambda x_1 + (1 - \lambda)x_2) \leq f(x_2)$. By differentiability of f at x_2 we have $f(\lambda x_1 + (1 - \lambda)x_2) = f(x_2) + \lambda\langle x_1 - x_2, \nabla f(x_2)\rangle + \lambda\|x_1 - x_2\| \cdot \epsilon(\lambda(x_1 - x_2))$ where $\epsilon(t) \to 0$ as $t \to 0$. Therefore $\lambda\langle x_1 - x_2, \nabla f(x_2)\rangle + \lambda\|x_1 - x_2\| \cdot \epsilon(\lambda(x_1 - x_2)) \leq 0$ for each $\lambda \in (0,1)$. Dividing by $\lambda > 0$ and letting $\lambda \to 0^+$ we get $\langle x_1 - x_2, \nabla f(x_2)\rangle$.

Conversely, suppose that $x_1, x_2 \in S$ with $f(x_1) \leq f(x_2)$ implies that $\langle x_1 - x_2, \nabla f(x_2)\rangle \leq 0$. We want to show that f is quasi-convex on S. Consider $x_1, x_2 \in S$ with $f(x_1) \leq f(x_2)$ and let $L = \{x : x = \lambda x_1 + (1 - \lambda)x_2, \lambda \in (0,1), f(x) > f(x_2)\}$. Note that the theorem is proved if we can show that $L = \emptyset$. We will first show that $\langle x_2 - x_1, \nabla f(x)\rangle = 0$ for each $x \in L$. We will then complete the proof by showing that under the assumption that $L \neq \emptyset$ we can construct a point $\bar{x} \in L$ with the property $\langle x_2 - x_1, \nabla f(\bar{x})\rangle \neq 0$.

Let $x = \lambda x_1 + (1 - \lambda)x_2 \in L$. Since $f(x_1) < f(x)$ then by hypothesis $0 \geq \langle x_1 - x, \nabla f(x)\rangle = (1 - \lambda)\langle x_1 - x_2, \nabla f(x)\rangle$ implying that $\langle x_1 - x_2, \nabla f(x)\rangle \leq 0$. Similarly $f(x_2) < f(x)$ implies that $\langle x_1 - x_2, \nabla f(x)\rangle \geq 0$ and hence we conclude that $\langle x_1 - x_2, \nabla f(x)\rangle = 0$ for each $x \in L$. Assume that $L \neq \emptyset$, i.e., there is an $\hat{x} \in L$. This means that $f(\hat{x}) > f(x_2) \geq f(x_1)$ and $\hat{x} = \lambda x_1 + (1 - \lambda)x_2$ for some $\lambda \in (0,1)$. Now since f is differentiable then f is also continuous. Therefore $f(\hat{x}) > f(x_2)$ implies that there is a $\delta \in [0,1)$ such that $f(\mu\hat{x} + (1 - \mu)x_2) > f(x_2)$ for each $\mu \in (\delta,1]$ with equality holding for $\mu = \delta$. Therefore by the mean value theorem

$$0 < f(\hat{x}) - f(\delta\hat{x} + (1 - \delta)x_2) = (1 - \delta)\langle\hat{x} - x_2, \nabla f(\bar{x})\rangle$$

for some $\bar{x} = \mu\hat{x} + (1 - \mu)x_2$ with $\mu \in (\delta,1)$. This means that $\bar{x} \in L$ and by dividing the above inequality by $1 - \delta$ we get $\langle \hat{x} - x_2, \nabla f(\bar{x}) \rangle > 0$ which in turn implies that $\langle x_1 - x_2, \nabla f(\bar{x}) \rangle > 0$. This contradicts the requirement $\langle x_1 - x_2, \nabla f(\bar{x}) \rangle = 0$ since $\bar{x} \in L$ and the proof is complete.

Another important class of functions is the class of pseudo-convex functions. Pseudo-convex functions are required to be differentiable. It may be noted that a differentiable convex function is indeed pseudo-convex. And in turn a pseudo-convex function is indeed quasi-convex and also strictly quasi-convex. This fact will be shown later.

4.5.6 Definition. Let S be an open set in E_n and let $f : S \to E_1$ be differentiable on S. f is said to be pseudo-convex on S if for each $x_1, x_2 \in S$ with $\langle x_1 - x_2, \nabla f(x_2) \rangle \geq 0$ we must have $f(x_1) \geq f(x_2)$. f is said to be strictly pseudo-convex on S if $x_1, x_2 \in S$ with $x_1 \neq x_2$ and $\langle x_1 - x_2, \nabla f(x_2) \rangle \geq 0$ imply that $f(x_1) > f(x_2)$.

Note that pseudo-convexity can be equivalently defined by: f is pseudo-convex on S if for each $x_1, x_2 \in S$, $f(x_1) < f(x_2)$ implies $\langle x_1 - x_2, \nabla f(x_2) \rangle < 0$.

f is said to be pseudo-concave (strictly pseudo-concave) if - f is pseudo-convex (strictly pseudo-convex). In other words f is pseudo concave if for each $x_1, x_2 \in S$ with $\langle x_1 - x_2, \nabla f(x_2) \rangle \leq 0$ implies that $f(x_1) \leq f(x_2)$. f is strictly pseudo-concave if $x_1, x_2 \in S$ with $x_1 \neq x_2$ and $\langle x_1 - x_2, \nabla f(x_2) \rangle \leq 0$ imply that $f(x_1) < f(x_2)$.

Note that in the above definition differentiability of f is assumed. So whenever we talk about pseudo-convex functions we implicitly assume differentiability of f. Also by the differentiability of f we implicitly assume that S is open. Now let f be a pseudo-convex function on S where $\nabla f(x_0) = 0$ and $x_0 \in S$. Then f has a global minimum over S at x_0. If f is strictly pseudo-concave then $\nabla f(x_0) = 0$ implies that f has a unique minimum over S at x_0.

An example of a pseudo-convex function is f defined by $f(x) = x^3 + x$ on E_1. Let $x_1, x_2 \in E_1$ be such that $\langle x_1 - x_2, \nabla f(x_2) \rangle \geq 0$, i.e., $\langle x_1 - x_2, 3x_2^2 + 1 \rangle \geq 0$. This implies that $x_1 \geq x_2$ and so $f(x_1) \geq f(x_2)$. This shows that f is indeed pseudo-convex.

The following theorem shows that a pseudo-convex function is both quasi-convex and strictly quasi-convex.

4.5.7 Theorem. Let S be an open convex set in E_n and let $f : S \to E_1$ be pseudo-convex (differentiable) on S. Then f is strictly quasi-convex and also quasi-convex on S.

Proof: Let f be pseudo-convex on S and suppose f is not strictly quasi-convex on S. We will show that this leads to a contradiction. If f is not strictly quasi-convex on S, there exist x_1 and $x_2 \in S$ with $f(x_2) < f(x_1)$ such that the set $L = \{x : x = \lambda x_1 + (1 - \lambda)x_2, \lambda \in (0,1), f(x) \geq f(x_1)\}$ is nonempty. Let $\hat{x} = \hat{\lambda} x_1 + (1 - \hat{\lambda})x_2 \in L$ for some $\hat{\lambda} \in (0,1)$. Since $f(\hat{x}) \geq f(x_1) > f(x_2)$ by the equivalent definition of pseudo-convexity, we have $\langle x_2 - \hat{x}, \nabla f(\hat{x})\rangle < 0$. Substituting $\hat{x} = \hat{\lambda}x_1 + (1 - \hat{\lambda})x_2$, this can be written as $\hat{\lambda}\langle x_2 - x_1, \nabla f(\hat{x})\rangle < 0$ which implies $\langle x_2 - x_1, \nabla f(\hat{x})\rangle < 0$. This in turn implies the existence of an $\bar{x} = \bar{\lambda}\hat{x} + (1 - \bar{\lambda})x_2$ with $\bar{\lambda} \in (0,1)$ such that $f(\bar{x}) > f(\hat{x}) \geq f(x_2)$. By pseudo-convexity of f we must then have $\langle \hat{x} - \bar{x}, \nabla f(\bar{x})\rangle < 0$ and $\langle x_2 - \bar{x}, \nabla f(\bar{x})\rangle < 0$. These two inequalities are contradicting by noting that $\bar{x} = \bar{\lambda}\hat{x} + (1 - \bar{\lambda})x_2$ and hence $x_2 - \bar{x} = -\frac{\bar{\lambda}}{1-\bar{\lambda}}(\hat{x} - \bar{x})$. Hence L is empty and f is strictly quasi-convex. To complete the proof, we note that f is lower semi-continuous since f is differentiable and from Theorem 4.5.3 f is quasi-convex.

The converse of the above theorem is not true. For example, $f(x) = x^3$ is strictly quasi-convex but not pseudo-convex.

At this stage we present another generalization of convexity. So far we have discussed convexity, quasi-convexity, and pseudo-convexity of a function on a given set. These concepts can be generalized to convexity, quasi-convexity, and pseudo-convexity at a point. This concept is useful because in many cases global convexity is really not needed for certain results in nonlinear programming to hold, but rather convexity at a point. We will give definitions corresponding to convexity at a point below. We will then give some of the important results regarding convexity at a point without any proofs. The proofs are actually very similar to other proofs in the case of global convexity discussed earlier in this chapter and hence are omitted.

Let $f : S \to E_1$ where S is a nonempty set of E_n. f is said to be <u>convex at $x_0 \in S$</u>

if $f(\lambda x_o + (1-\lambda)x) \leq \lambda f(x_o) + (1-\lambda)f(x)$ for each $x \in S$ and $\lambda \in (0,1)$ with $\lambda x_o + (1-\lambda)x \in S$. f is said to be <u>quasi-convex at x_o</u> $\in S$ if for each $x \in S$ with $f(x) \leq f(x_o)$ it follows that $f(\lambda x_o + (1-\lambda)x) \leq f(x_o)$ whenever $\lambda x_o + (1-\lambda)x \in S$ with $\lambda \in (0,1)$. f is said to be <u>pseudo-convex at x_o</u> if $\langle x - x_o, \nabla f(x_o)\rangle \geq 0$ implies that $f(x) \geq f(x_o)$ for each $x \in S$.

Based on the above definitions, strict convexity, strict quasi-convexity, and strict pseudo-convexity at a point can be given. Also the cases of different types of concavity and strict concavity at a point can be similarly defined.

The following results can be easily checked.

1. $f : S \to E_1$ is convex at $x_o \in S$ if and only if E_f is convex (star shaped) at $(x_o, f(x_o))$.
2. Let $f : S \to E_1$ be differentiable and convex at $x_o \in \text{int } S$. Then $f(x) \geq f(x_o) + \langle x - x_o, \nabla f(x_o)\rangle$ for each $x \in S$. If f is strictly convex at $x_o \in \text{int } S$ then $f(x) > f(x_o) + \langle x - x_o, \nabla f(x_o)\rangle$ for each $x \in S$ with $x \neq x_o$.
3. Let $f : S \to E_1$ be twice differentiable at and convex at $x_o \in \text{int } S$. Then the Hessian matrix of f at x_o is positive semi-definite.
4. Let $f : S \to E_1$ be differentiable and quasi-convex at $x_o \in \text{int } S$. Then $x \in S$ with $f(x) \leq f(x_o)$ implies that $\langle x - x_o, \nabla f(x_o)\rangle \leq 0$.

CHAPTER 5

STATIONARY POINT OPTIMALITY CONDITIONS WITH DIFFERENTIABILITY

In this chapter we will be concerned with developing optimality conditions for a nonlinear programming problem. We will consider different problems of varying difficulties and develop necessary conditions that a point must satisfy in order to qualify for an optimal solution. Under certain convexity assumptions we also show that these conditions are sufficient for optimality. Unlike the saddle conditions of Chapter 7, the optimality criteria of this chapter use the gradients of the objective function and constraint functions.

5.1. Inequality Constrained Problems

In this section we consider a problem of minimizing an objective function over a constraint set. The constraint set is defined via an arbitrary set as well as constraints of the inequality type. We require the functions involved to be differentiable at the point under consideration. To be more precise, we consider the following problem P.

Problem P: Minimize $f(x)$ subject to $x \in X$ and $g_i(x) \leq 0$, $i = 1,2,\ldots,m$. If we denote $(g_1,g_2,\ldots,g_m)$ by g and let $S = \{x \in X : g(x) \leq 0\}$ then the problem can be restated as: minimize $\{f(x) : x \in S\}$. Here $f,g_1,g_2,\ldots,g_m$ are defined on an open set containing X and take values in E_1. S is the feasible set (region) and any $x \in S$ is said to be a feasible solution of problem P.

Now suppose that x_o is an optimal solution of problem P above, i.e., $x \in S$ implies that $f(x_o) \leq f(x)$. Of course we are implicitly assuming that x_o is a feasible solution of P. The purpose here is to develop some conditions which must hold if x_o is an optimal solution of problem P. These conditions are called necessary conditions for optimality. The reader may note that the statement x_o solves (or is an optimal solution of) problem P is equivalent to a certain set being empty, namely $\{x \in S : f(x) - f(x_o) < 0\}$. In other words optimality of x_o is essentially equivalent to $F \cap X \cap G = \emptyset$ where $F = \{x : f(x) < f(x_o)\}$, $G = \bigcap_{i=1}^{m} G_i$, $G_i = \{x : g_i(x) \leq 0\}$, $i = 1,2,\ldots,m$.

It will turn out that the necessary conditions we seek follow from invoking separation theorems of disjoint sets. However, since the above disjoint sets are not convex, we first need to make them convex (approximate them by convex sets) in such a way to keep their disjointness. Once this is done, we then can use separation theorems of nonintersecting convex sets. In other words, our main objective here is to replace the above disjoint sets by another set of disjoint sets which are convex and suitably related to the above sets.

Now let $x_0 \in \text{int } X$. We will use the following notation. $F' = \{x : \langle x, \nabla f(x_0)\rangle < 0\}$, $G_i' = \{x : \langle x, \nabla g_i(x_0)\rangle < 0\}$, $G' = \bigcap_{i \in I} G_i'$ where $I = \{i : g_i(x_0) = 0\}$, i.e., I is the set of binding (effective) constraints at the point x_0. We will use F' and G' to approximate F and G. In other words the convex approximations are the open halfspace corresponding to the inequality constraints and the objective function. Note that we do not explicitly introduce a suitable convex approximation to X because the suitable convex approximation to X corresponding to $x_0 \in \text{int } X$ is E_n itself. This will be clear later when we let the convex approximation of X at x_0 be the cone of interior directions (see Section 3.5.7) which is indeed E_n if $x_0 \in \text{int } X$. The following theorem is essential in proving the Fritz John conditions to be developed later.

5.1.1 <u>Theorem</u>. Suppose that $x_0 \in \text{int } X$ solves problem P and suppose that f, and $g_i (i \in I)$ are differentiable at x_0 and g_i be continuous at x_0 for $i \notin I$. Then $F' \cap G' = \emptyset$.

<u>Proof</u>: We will show that if $F' \cap G'$ is not empty then a feasible point z can be constructed with $f(z) < f(x_0)$, violating optimality of x_0. Now let $x \in F' \cap G'$. Since $x_0 \in \text{int } X$ and $g_i(x_o) < 0$ for $i \notin I$, then $x_0 + \lambda x \in X$ and $g_i(x_0 + \lambda x) \leq 0$ for each $i \notin I$, for λ sufficiently small. We now show that $x_0 + \lambda x$ is indeed feasible to the problem for $\lambda > 0$ sufficiently small. For $i \in I$, $g_i(x_0 + \lambda x) = g_i(x_0) + \lambda\langle x, \nabla g_i(x_0)\rangle + \lambda\|x\| \cdot \epsilon_i(\lambda x)$ by differentiability of g_i at x_0. But since $x \in G_i'$ for $i \in I$ then $\langle x, \nabla g_i(x_0)\rangle < 0$ and so for $\lambda > 0$ sufficiently small we must have $g_i(x_0 + \lambda x) < 0$. This is due to the fact that $\epsilon_i(\lambda x) \to 0$ as $\lambda \to 0$ and so $\lambda\langle x, \nabla g_i (x_0)\rangle$ dominates $\lambda\|x\| \cdot \epsilon_i(\lambda x)$ for λ sufficiently small. This shows that $x_0 + \lambda x \in S$ for $\lambda > 0$ sufficiently small. But on the other hand $\langle x, \nabla f(x_0)\rangle < 0$ implies that

$f(x_o + \lambda x) = f(x_o) + \lambda\langle x, \nabla f(x_o)\rangle + \lambda\|x\|\cdot \epsilon(\lambda x) < f(x_o)$ for $\lambda > 0$ sufficiently small. This provides feasible points $x_o + \lambda x$ with $f(x_o + \lambda x) < f(x_o)$, which is impossible in view of optimality of x_o. So $F' \cap G' = \emptyset$.

Let us discuss the implications of the above theorem. If x_o solves the problem then $F' \cap G' = \emptyset$, i.e., we can find no vector x such that $\langle x, \nabla f(x_o)\rangle < 0$ and meanwhile $\langle x, \nabla g_i(x_o)\rangle < 0$ for each $i \in I$.

Geometrically the condition $F' \cap G' = \emptyset$ means that there is no halfspace that contains the gradient vectors $\nabla f(x_o)$ and $\nabla g_i(x_o)$ for $i \in I$. This also implies that we can find a nontrivial (i.e., nonzero) nonnegative combination of these vectors which add up to zero. In other words we can find numbers u_o, $u_i (i \in I)$ which are nonnegative and not all are zeros such that $u_o \nabla f(x_o) + \sum_{i \in I} u_i \nabla g_i(x_o) = 0$. These are precisely the Fritz John necessary conditions for optimality. This discussion is made more precise by Theorem 5.1.3 below. First the following lemma is needed.

5.1.2 Lemma. Suppose that the system $Ax \in K$ where $x \in S$ is inconsistent. Here A is an mxn matrix, K is a convex cone in E_m and S is a convex set in E_n. Then there exists a nonzero $p \in K^*$ such that $\langle p, Ax\rangle \geq 0$ for each $x \in S$.

Proof: The set $A(S) = \{Ax : x \in S\}$ is convex and by hypothesis $A(S) \cap K$ is empty. By Theorem 2.4.14 there is a nonzero p such that $\langle p, Ax\rangle \geq \langle p, y\rangle$ for each $x \in S$ and $y \in K$. It is clear that the above inequality holds for $y = 0 \in C\ell\, K$ and so $\langle p, Ax\rangle > 0$ for each $x \in S$. It is also clear that $\langle p, y\rangle \leq 0$ for every $y \in K$ since otherwise there must exist a $y_o \in K$ with $\langle p, y_o\rangle > 0$ and then $\langle p, Ax\rangle \geq \lambda\langle p, y_o\rangle$ for every $\lambda > 0$, which is impossible. This shows that $p \in K^*$ and the proof is complete.

Corollary. If $S = E_n$ then $A'p = 0$.

Proof: Note that $\langle p, Ax\rangle \geq 0$ for every $x \in E_n$ and hence $\langle x, A'p\rangle \geq 0$ for every $x \in E_n$. Letting $x = -A'p$ it follows that $A'p = 0$.

We will make use of the above lemma to prove the following theorem. Actually in Theorem 5.1.3 below K is the negative orthant. We choose to put Lemma 5.1.2 above in a more general form because different appropriate cones K will be specified later for different problems, e.g., problems with equality and inequality constraints.

5.1.3 Theorem. Suppose that $x_o \in \text{int}\, X$ solves problem P and let f and $g_i (i \in I)$ be differentiable at x_o. Let g_i be continuous at x_o for $i \notin I$. Then there exist nonnegative numbers u_o, $u_i (i \in I)$, not all zero, such that $u_o \nabla f(x_o) + \sum_{i \in I} u_i \nabla g_i(x_o) = 0$.

<u>Proof</u>: Since x_o solves P then by Theorem 5.1.1 we have $F' \cap G' = \emptyset$, i.e., the system $Ax < 0$ is inconsistent, where A is the matrix formed by $\nabla f(x_o)$, and $\nabla g_i(x_o)$ for $i \in I$ as its columns. Now apply Lemma 5.1.2 above where K is the negative orthant. Therefore there must exist a nonzero vector $p \in K^*$ such that $A'p = 0$. Noting that K^* is the nonnegative orthant it is clear that $p \geq 0$. Therefore there must exist nonnegative scalars u_o and $u_i (i \in I)$ such that:

i. Not all u_o and $u_i (i \in I)$ are zero

ii. $u_o \nabla f(x_o) + \sum_{i \in I} u_i \nabla g_i(x_o) = 0$

Here u_o and u_i's are the components of the vector p. This completes the proof.

The above conditions are very important. Any point that does not satisfy these conditions cannot be a candidate for an optimal solution of the nonlinear programming problem P. These conditions are referred to as the Fritz John conditions. Now if $g_i (i \notin I)$ are also differentiable at x_o, then we can write the Fritz John conditions as follows. If x_o solves the problem then there must exist a vector $(u_o, u_1, \ldots, u_m)$ such that:

i. $u_o \geq 0$, $u_i \geq 0$ for $i = 1,2,\ldots,m$.

ii. $(u_o, u_1, \ldots, u_m)$ is not the zero vector in E_{m+1}

iii. $u_i g_i(x_o) = 0$ for $i = 1,2,\ldots,m$.

iv. $u_o \nabla f(x_o) + \sum_{i=1}^{m} u_i \nabla g_i(x_o) = 0$

The equivalence between the two forms of the optimality conditions should be clear. In the form mentioned above we merely let $u_i = 0$ for $i \notin I$. Note that iii follows since for $i \in I$, $g_i(x_o) = 0$ and for $i \notin I, u_i = 0$.

u_o is called the Lagrangian multiplier corresponding to the objective function while u_i is the Lagrangian multiplier corresponding to constraint i. Note that the Lagrangian multipliers corresponding to inactive (nonbinding) constraints are zeros.

Usually condition (iii) above is referred to as complementarity slackness condition.

It should be noted that the above conditions (Fritz John conditions) are necessary for optimality, but in general are not sufficient, i.e., if a point satisfies the above conditions we may or may not have an optimal point.

It may be helpful to give some examples of optimal points and to develop their corresponding Fritz John conditions. Figure 5.1 shows two examples. Figure 5.1(a) corresponds to the problem of minimizing $f(x,y) = -x$, subject to $g_1(x,y) = -y \leq 0$ and $g_2(x,y) = -(1-x)^3 + y \leq 0$. It is clear that the optimal point is (1,0) since any point (x,y) with $x > 1$ must have $y < 0$ in order to satisfy $g_2(x,y) \leq 0$. But this violates $g_1(x,y) \leq 0$. Now $\nabla f(1,0) = (-1,0)$, $\nabla g_1(1,0) = (0,-1)$, and $\nabla g_2(x,y) = (0,1)$. So the Fritz John conditions hold with $u_o = 0$, $u_1 = 1$, and $u_2 = 1$. Figure 5.1(b) corresponds to the problem of minimizing the same objective function. Here the second constraint is replaced by $g_2(x,y) = x + y - 1 \leq 0$. The optimal point is still (1,0). As before $\nabla f(1,0) = (-1,0)$ and $\nabla g_1(1,0) = (0,-1)$. Now $\nabla g_2(1,0) = (1,1)$ and so the Fritz John conditions hold with $u_o = u_1 = u_2 = 1$.

From the above example we can see two possible cases. The first case corresponds to vanishing Lagrangian multiplier of the objective function and the second case corresponds to nonvanishing Lagrangian multipliers. In the second case the conditions are referred to as the Kuhn-Tucker conditions. Actually since $u_o > 0$ then one can divide the Fritz John equation by u_o and get some new Lagrangian multipliers. In other words the Kuhn-Tucker conditions read as follows. If x_o solves problem P then there is a vector $(u_1, u_2, \ldots, u_m)$ such that:

i. $u_i \geq 0 \quad i = 1,2,\ldots,m$

ii. $u_i g_i(x_o) = 0 \quad i = 1,2,\ldots,m$

iii. $\nabla f(x_o) + \sum_{i=1}^{m} u_i \nabla g_i(x_o) = 0$

Clearly one may restrict the attention to binding constraints only. In this case we must have $u_i \geq 0$ for each $i \in I$ and $\nabla f(x_o) + \sum_{i \in I} u_i \nabla g_i(x_o) = 0$.

We will show that under some convexity assumptions the Kuhn-Tucker conditions are sufficient for optimality. This makes it important to see under what conditions we are assured of positivity of the Langrangian multiplier of the objective

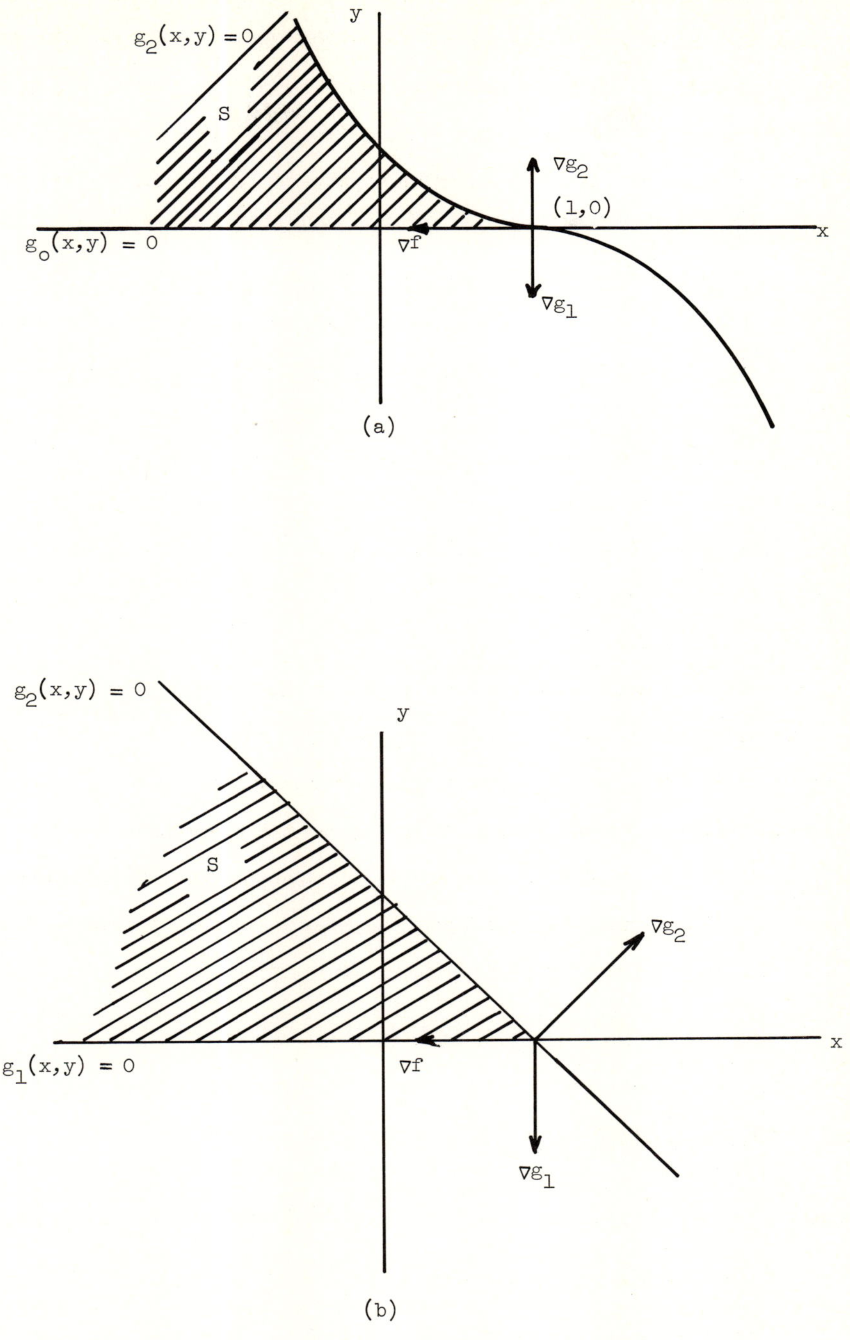

<u>Figure 5.1</u>

<u>Illustrative Examples of the Fritz John Conditions</u>

function. Conditions of this nature are usually referred to as constraint qualifications since they involve the constraints only. Constraint qualifications are discussed in detail later in Chapter 6. However we give below a simple constraint qualification which guarantees that $u_o > 0$ and hence the Kuhn-Tucker conditions.

Suppose that $x_o \in \text{int } X$ solves problem P. Then by the Fritz John conditions there exist a nonzero nonnegative vector (u_o, u_I) such that $u_o \nabla f(x_o) + \sum_{i \in I} u_i \nabla g_i(x_o) = 0$ where $I = \{i : g_i(x_o) = 0\}$ and u_I is the vector with components $u_i (i \in I)$.

Now let us pose the following <u>constraint qualification</u>. Suppose that $\nabla g_i(x_o)$ for $i \in I$ are linearly independent. Then it is obvious that $u_o > 0$ because otherwise $u_o = 0$ and we must have $\sum_{i \in I} u_i \nabla g_i(x_o) = 0$ and $u_I \neq 0$, violating the independence assumption. Therefore $u_o > 0$ and without loss of generality we let $u_o = 1$, which leads to the Kuhn-Tucker conditions. The reader may note that in the example corresponding to Figure 5.1(a) above $\nabla g_1(x_o)$ and $\nabla g_2(x_o)$ are not linearly independent and the Lagrangian multiplier corresponding to the objective function is zero.

The following theorem shows that under moderate convexity assumptions, the Kuhn-Tucker conditions are also sufficient for optimality.

5.1.3 <u>Theorem</u>. Let $x_o \in S$ and suppose that f is pseudo-convex at x_o and $g_i (i \in I)$ be quasi-convex at x_o, where $I = \{i : g_i(x_o) = 0\}$. Suppose that the Kuhn-Tucker conditions hold at x_o, i.e., there exist nonnegative scalars $u_i (i \in I)$ such that $\nabla f(x_o) + \sum_{i \in I} u_i \nabla g_i(x_o) = 0$. Then x_o solves problem P.

<u>Proof</u>: Let $x \in S$. We will first show that $\langle x - x_o, \nabla f(x_o) \rangle \geq 0$. For $i \in I$, $g_i(x) \leq 0$ and $g_i(x_o) = 0$, i.e., $g_i(x) \leq g_i(x_o)$. By quasi-convexity of g_i at x_o it follows that $\langle x - x_o, \nabla g_i(x_o) \rangle \leq 0$. Multiplying by u_i and summing over I we get $\langle x - x_o, \sum_{i \in I} u_i \nabla g_i(x_o) \rangle \leq 0$. But since $\nabla f(x_o) + \sum_{i \in I} u_i \nabla g_i(x_o) = 0$ it follows that $\langle x - x_o, \nabla f(x_o) \rangle \geq 0$. Now the result is obvious by pseudo-convexity of f at x_o.

Needless to say that if f and g_i are convex at x_o (and hence both pseudo-convex and quasi-convex at x_o) then the Kuhn-Tucker conditions are sufficient. Also if convexity at a point is replaced by the stronger requirement of global convexity the

Kuhn-Tucker conditions are also sufficient. As mentioned earlier the Kuhn-Tucker conditions are necessary for optimality provided that any of the constraint qualifications (to be discussed in the next chapter) hold, even in the absence of convexity.

5.2. Equality and Inequality Constrained Constraints

In this section we want to consider a more general problem which has both equality and inequality constraints. Our objective is to develop necessary conditions for optimality. Here we consider the problem P to minimize $f(x)$ subject to $x \in X$, $g(x) \le 0$, and $h(x) = 0$. Here $g = (g_1, g_2, \ldots, g_m)$ and $h = (h_1, h_2, \ldots, h_k)$. As seen below, the presence of the equality constraints considerably complicates the problem.

The question that arises is: can we transform this problem into a problem of the inequality type and use the material developed in the previous section? One may be tempted to replace each inequality constraint $h_i(x) = 0$ by two inequality restrictions, namely $h_i(x) \le 0$ and $-h_i(x) \le 0$. However this does not solve the problem because the Fritz John conditions become useless and every point will satisfy these conditions. To clarify, replace the equality constraints by inequality restrictions as mentioned above to get the problem: minimize $f(x)$ subject to $g_i(x) \le 0$ for $i = 1,2,\ldots,m$, $h_i(x) \le 0$, $-h_i(x) \le 0$ for $i = 1,2,\ldots,k$. Now let x be any feasible point. It is clear that $u_0 \nabla f(x_0) + \sum_{i=1}^{m} u_i \nabla g_i(x) + \sum_{i=1}^{k} v_i \nabla h_i(x)$

$$+ \sum_{i=1}^{k} w_i \nabla(-h_i(x)) = 0$$ where $u_0 = u_i = 0$ for each $i = 1,2,\ldots,m$ and $v_i = 1$ for each $i = 1,2,\ldots,k$. Note that $u_i g_i(x) = v_i h_i(x) = w_i h_i(x) = 0$. This shows that each feasible point will trivially satisfy the above conditions which then become useless.

In the discussion that follows we will develop Fritz John type conditions for problems with equality and inequality restrictions. As in Section 1 the key idea is to convert the statement x_0 solves P into disjoint convex sets using suitable convex approximations of the sets under consideration. For an equality constraint a reasonable approximation is the tangent hyperplane $\{x : \langle x - x_0, \nabla h(x_0)\rangle = 0\}$ to the set $\{x : h(x) = 0\}$. However, the main idea in proving disjointness of the approximating sets was to find a feasible point with better objective function if there is one in the intersection of these sets, to verify the optimality assumption of the point under consideration (see Theorem 5.1.1). However, we cannot use the same argument as before because if we move in the direction of any point in the approximation of the equality constraints we will leave the feasible set. In developing the Fritz John conditions we will still use the same general strategy with a modification which insures that we do not leave the feasible set.

We will use the following notation in Theorem 5.2.1 below. Let F and G be as defined before, i.e., $F = \{x : f(x) < f(x_o)\}$, $G = \bigcap_{i=1}^{m} G_i$ where $G_i = \{x : g_i(x) \leq 0\}$. Also let $H = \bigcap_{i=1}^{k} H_i$ where $H_i = \{x : h_i(x) = 0\}$. Note that the statement that $x_o \in S$ solves problem P is equivalent to saying that $F \cap G \cap H = \emptyset$. Now let $F' = \{x : \langle x, \nabla f(x_o)\rangle < 0\}$, $G' = \bigcap_{i \in I} G_i'$, $G_i' = \{x : \langle x, \nabla g_i(x_o)\rangle < 0\}$ and

$H' = \bigcap_{i=1}^{k} H_i'$, $H_i' = \{x : \langle x, \nabla h_i(x_o)\rangle = 0\}$. The following theorem essentially replaces disjointness of F, G, and H by disjointness of the convex sets F', G', and H'. In the theorem the following notation is used. Let M be the nxk matrix formed by $\nabla h_1(x_o), \nabla h_2(x_o), \ldots, \nabla h_k(x_o)$ as its columns. Clearly H' is the null space of the matrix M. We let P_o be the projection matrix into H', i.e., $P_o = I - M^t(MM^t)^{-1}M$. Note that for any vector x, $P_o x \in H'$ because $MP_o x = M[x - M^t(MM^t)^{-1}Mx] = 0$. By the decomposition Theorem 1.3.3 it is clear that $P_o x$ is the unique projection into H' and $M^t(MM^t)^{-1}Mx$ is the projection on the orthogonal complement of H'.

5.2.1 <u>Theorem</u>. Let $x_o \in \text{int } X$ solve problem P. Suppose that $f, g_i (i \in I)$ be differentiable at x_o and $g_i (i \notin I)$ be continuous at x_o. Suppose that $h_i (i = 1,2,\ldots,k)$ be continuously differentiable at x_o and $\nabla h_1(x_o), \ldots, \nabla h_k(x_o)$ are linearly independent. Then $F' \cap G' \cap H' = \emptyset$.

<u>Proof</u>: By contradiction, suppose that $F' \cap G' \cap H' \neq \emptyset$, i.e., we assume that there is an element y in $F' \cap G' \cap H'$. We will construct a function $\alpha : E_1 \to E_n$ corresponding to a feasible arc with $f(\alpha(\theta)) < f(x_o)$ for each $\theta \in (0, \theta_o]$ for some $\theta_o > 0$, violating optimality of x_o. For any $\theta > 0$ let M_θ be the nxk matrix formed by $\nabla h_1(\alpha(\theta))$, $\nabla h_2(\alpha(\theta)), \ldots, \nabla h_k(\alpha(\theta))$ and let P_θ be the matrix which projects any x into the null space of M_θ. We then define the arc by the following differential equation: $\alpha(0) = x_o$ and $\frac{d}{d\theta}\alpha(\theta) = P_\theta y$. This equation is well defined for $\theta > 0$

sufficiently small because of linear independence of $\nabla h_1(x_o),\ldots,\nabla h_k(x_o)$ and continuous differentiability of $h_1,h_2,\ldots,h_k$ at x_o. We will first show that $g(\alpha(\theta)) \leq 0$ and $\alpha(\theta) \in X$ for sufficiently small $\theta > 0$. For each $i \in I$ we get by the chain rule:

$$\frac{d}{d\theta} g_i(\alpha(\theta))\Big|_{\theta = 0} = \langle \frac{d}{d\alpha(\theta)} g_i(\alpha(\theta)), \frac{d\alpha(\theta)}{d\theta}\rangle\Big|_{\theta = 0}$$

$$= \langle \nabla g_i(\alpha(\theta)), P_\theta y \Big|_{\theta = 0}$$

$$= \langle \nabla g_i(x_o), P_o y\rangle .$$

But since $y \in H'$ then $P_o y = y$ and hence $\frac{d}{d\theta} g_i(\alpha(0)) = \langle \nabla g_i(x_o), y\rangle < 0$ since $y \in G_i'$. This and differentiability of g_i at x_o then implies that for each $i \in I$ there is a $\theta_i > 0$ such that $g_i(\alpha(\theta)) \leq 0$ for all $\theta \in [0,\theta_i]$. Let $\epsilon_1 = \min_{i \in I} \theta_i$. By continuity of g_i at x_o and since $g_i(x_o) < 0$ for $i \notin I$ it is clear that there is an $\epsilon_2 > 0$ such that $g_i(\alpha(\theta)) \leq 0$ for each $\theta \in [0,\epsilon_2]$ for each $i \notin I$. Also since $x_o \in \text{int } X$, there is an $\epsilon_3 > 0$ such that $\alpha(\theta) \in X$ for each $\theta \in [0,\epsilon_3]$. Now let $\theta_o = \min(\epsilon_1,\epsilon_2,\epsilon_3)$. Then obviously $\alpha(\theta) \in X$ and $g(\alpha(\theta)) \leq 0$ for each $\theta \in [0,\theta_o]$. To show feasibility of $\alpha(\theta)$ we need to show that $h_i(\alpha(\theta)) = 0$ for $i = 1,2,\ldots,k$. But $h_i(\alpha(\theta)) = h_i(\alpha(0)) + \theta \frac{d}{d\mu} h_i(\alpha(\mu))$ for some $\mu \in [0,\theta]$. Note that

$$\frac{d}{d\mu} h_i(\alpha(\mu)) = \langle\frac{d}{d\alpha(\mu)} h_i(\alpha(\mu)), \frac{d\alpha(\mu)}{d\mu}\rangle$$

$$= \langle \nabla h_i(\alpha(\mu)), P_\mu y\rangle .$$

But it is clear that $P_\mu y$ is in the null space of M_μ and $\nabla h_i(\alpha(\mu))$ is in the range space of M_μ and so $\langle \nabla h_i(\alpha(\mu)), P_\mu y\rangle = 0$. This shows that $\frac{d}{d\mu} h_i(\alpha(\mu)) = 0$ and hence $h_i(\alpha(\theta)) = 0$. This then shows that for each $\theta \in [0,\epsilon_o], \alpha(\theta) \in S$, i.e., the arc we constructed is indeed feasible. We will show that the objective function decreases along this arc. Note that

$$\frac{d}{d\theta} f(\alpha(\theta))\Big|_{\theta = 0} = \langle\frac{d}{d\alpha(\theta)} f(\alpha(\theta)), \frac{d\alpha(\theta)}{d\theta}\rangle \Big|_{\theta = 0}$$

$$= \langle \nabla f(\alpha(\theta), P_\theta y\rangle\Big|_{\theta = 0}$$

has a solution $x \in E_n$. Moreover $\nabla h_1(x_o),\ldots,$ and $\nabla h_k(x_o)$ are linearly independent.

5.2.3 <u>Theorem</u>. (Kuhn-Tucker Conditions). Suppose that $x_o \in \text{int } X$ solves the problem: minimize $f(x)$ subject to $x \in X$, $g(x) \le 0$, $h(x) = 0$. Suppose that $\nabla h_1(x_o)$, $\nabla h_2(x_o),\ldots,\nabla h_k(x_o)$ are linearly independent and that there exists a vector x such that $\langle x,\nabla g_i(x_o)\rangle < 0$ for $i \in I$ and $\langle x,\nabla h_i(x_o)\rangle = 0$ for $i = 1,2,\ldots,k$. Then there must exist a vector (u_I,v) such that $u_i \ge 0$ for each $i \in I$, and $\nabla f(x_o) + \sum_{i \in I} u_i\nabla g_i(x_o) + \sum_{i=1}^{k} v_i\nabla h_i(x_o) = 0$.

<u>Proof</u>: By the Fritz John conditions there is a nonzero vector (u_o,u_I,v) with $u_o \ge 0$, $u_i \ge 0$ $(i \in I)$ such that $u_o\nabla f(x_o) + \sum_{i \in I} u_i\nabla g_i(x_o) + \sum_{i=1}^{k} v_i\nabla h_i(x_o) = 0$. We will show that $u_o > 0$. By contradiction suppose that $u_o = 0$. Therefore $\sum_{i \in I} u_i\nabla g_i(x_o) + \sum_{i=1}^{k} v_i\nabla h_i(x_o) = 0$. Multiplying by x and noting that $\langle x,\nabla h_i(x_o)\rangle = 0$ for $i = 1,2,\ldots,k$ it is clear that $\sum_{i \in I} u_i \langle x,\nabla g_i(x_o)\rangle = 0$. But since $\langle x,\nabla g_i(x_o)\rangle < 0$ and $u_i \ge 0$ for each $i \in I$ it follows that $u_i = 0$ for each $i \in I$. This shows that $\sum_{i=1}^{k} v_i\nabla h_i(x_o) = 0$. But since $u_o = 0$ and $u_i = 0$ for every $i \in I$ it follows that $(v_1,v_2,\ldots,v_k) \neq 0$. But this violates independence of $\nabla h_1(x_o),\ldots,$ $\nabla h_k(x_o)$. Therefore $u_o > 0$. Without loss of generality we can assume that $u_o = 1$ and the Kuhn-Tucker conditions follow.

As we have done earlier in the case of pure inequality constraints the Kuhn-Tucker conditions are sufficient for the mixed case under suitable convexity assumptions on the inequality constraint functions and linearity of the equality constraints. This is given by the following theorem.

5.2.4 <u>Theorem</u>. (Sufficiency of the Kuhn-Tucker Conditions). Consider the problem: minimize $f(x)$ subject to $x \in X$, $g(x) \le 0$ and $h(x) = 0$. Suppose that x_o is a feasible solution and let $I = \{i : g_i(x_o) = 0\}$. Suppose that $h_i (i = 1,2,\ldots,k)$ are linear, $g_i (i \in I)$ are quasi-convex at x_o, and f is pseudo-convex at x_o. If the Kuhn-Tucker conditions hold at x_o then x_o solves the problem.

Proof: By the Kuhn-Tucker conditions there exist $u_i \geq 0$ $(i \in I)$ and $v_i (i = 1,2,\ldots,k)$ such that $\nabla f(x_o) + \sum_{i \in I} u_i \nabla g_i(x_o) + \sum_{i=1}^{k} v_i \nabla h_i(x_o) = 0$. We will first show that any feasible x must satisfy $\langle x - x_o, \nabla f(x_o)\rangle \geq 0$. Let x be any feasible solution of the problem. Then for $i \in I$, $g_i(x) \leq 0$. By quasi-convexity of g_i at x_o we must have $\langle x - x_o, \nabla g_i(x_o)\rangle \leq 0$. By feasibility of x we must have $h_i(x) = 0$ and by linearity of h_i we must have $\langle x - x_o, \nabla h_i(x_o)\rangle = 0$ for $i = 1,2,\ldots,k$. Putting these facts together and noting that $u_i \geq 0$ we must have $\sum_{i \in I} u_i \langle x - x_o, \nabla g_i(x_o)\rangle + \sum_{i=1}^{k} v_i \langle x - x_o, \nabla h_i(x_o)\rangle \leq 0$. But noting that $\nabla f(x_o) + \sum_{i \in I} u_i \nabla g_i(x_o) + \sum_{i=1}^{k} v_i \nabla h_i(x_o) = 0$ we get $\langle x - x_o, \nabla f(x_o)\rangle \geq 0$. By pseudo-convexity of f at x_o it follows that $f(x) \geq f(x_o)$ and the proof is complete.

5.3. Optimality Criteria of the Minimum Principle Type

In the last two sections we had the requirement that $x_o \in \text{int } X$. If this assumption is relaxed then we cannot in general obtain the Fritz John conditions. Actually one obtains in this case conditions of the minimum principle type. To illustrate this point let us consider this simple example. We are trying to minimize $f(x,y) = -x + y$ subject to $h(x,y) = (x-1)^2 + y^2 - 1 = 0$. X is a nonfunctionally specified set which is shown in Figure 5.2 (X can also be developed by suitable inequality constraints). It is clear that the optimal point is $\left(\frac{\sqrt{2}-1}{\sqrt{2}}, \frac{-1}{\sqrt{2}}\right)$. We will show that the Fritz John conditions for optimality do not hold at this point. Here the problem is that the above point does not belong to the interior of X. Now $\nabla f(x_o,y_o) = (-1,1)$ and $\nabla h(x_o,y_o) = (-\sqrt{2}, -\sqrt{2})$.

If the Fritz John conditions of Theorem 5.2.2 were to hold then we must have a nonzero (u_o,v) with $u_o \geq 0$ such that $u_o \nabla f(x_o,y_o) + u\nabla h(x_o,y_o) = 0$. In other words $u_o(-1,1) + v(-\sqrt{2}, -\sqrt{2}) = (0,0)$. Solving these two equations we get $u_o = v = 0$. Put differently the Fritz John conditions for optimality do not hold at this point.

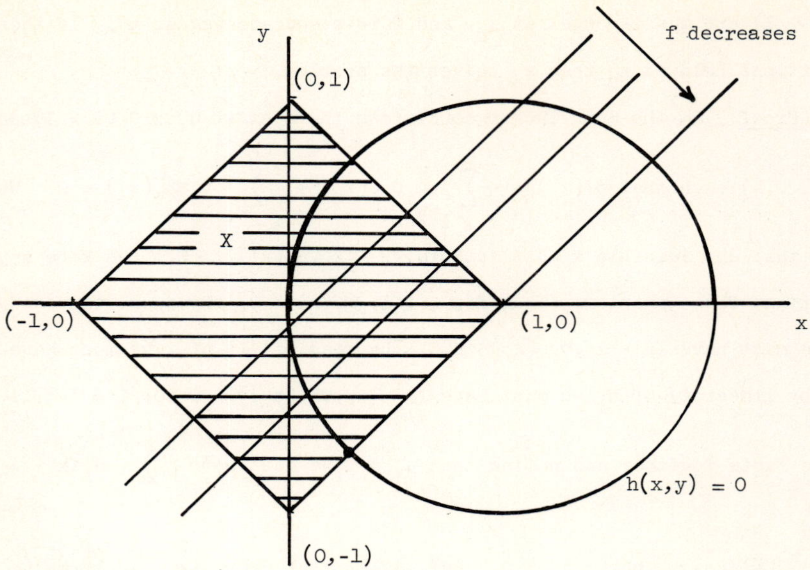

Figure 5.2

An Illustrative Example

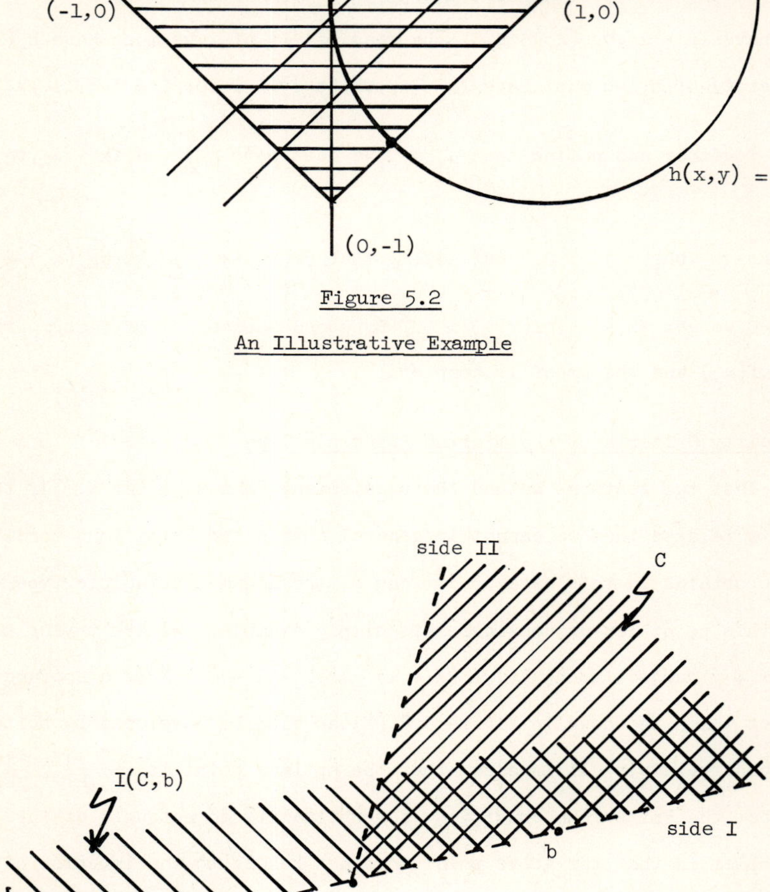

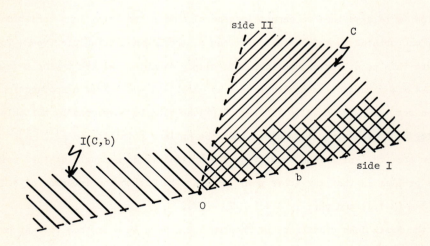

Figure 5.3

Cone of Interior Directions

In this section we will modify the Fritz John conditions to take care of the case when $x_o \notin \text{int } X$. The inequality constraints are defined via a convex cone and the equality constraints are singled out. As mentioned earlier the optimality criteria developed are of the minimum principle type which parallel the necessary conditions for optimal control problems.

In order to develop the necessary optimality conditions of this section two lemmas are needed. The first lemma below characterizes the cone of interior directions when the set under consideration is a convex cone. The second lemma replaces a nonintersecting collection of nonconvex sets by a collection of nonintersecting convex sets. These sets are the proper convex approximations (linearizations) of the original sets. Lemma 5.3.2 actually generalizes Theorems 5.1.1 and 5.1.2 given earlier.

5.3.1 Lemma. Let C be an open convex cone and $b \in C\ell\, C$. Let $C(b) = \{c - \alpha b :$ $c \in C, \alpha \geq 0\}$. Then $C(b) = I(C,b)$.

Proof: Let $a \in C(b)$. Then $a = c - \alpha b$ where $c \in C$ and $\alpha \geq 0$. Now let N be a ball about the origin such that $c + B \subset C$. Note that such a ball exists since C is open and $c \in C$. Let δ be a positive number such that $\alpha\delta < 1$. Then for $z \in N$, $\lambda \in (0,\delta)$ we have $b + \lambda(a + z) = (1 - \lambda\alpha)b + \lambda(c + z) \in C\ell\, C + C = C$ since C is an open convex cone. Thus $a \in I(c,b)$. Conversely let $a \in I(c,b)$. Then there exists a $\delta > 0$ such that for each $\lambda \in (0,\delta)$, $b + \lambda a \in C$. Since C is a cone then $a + \frac{1}{\lambda} b \in C$ and hence $a = c - \frac{1}{\lambda} b$ with $c \in C$. Hence $a \in C(b)$ and the proof is complete.

From the above it is clear that $C(b) = I(c,b)$ is an open convex cone which contains C. It will be clear later that using the cone of interior directions rather than the cone itself will give us the complementarity slackness conditions. This notion in some sense replaces the notion of binding constraints. As shown in Figure 5.3 side I of the cone is binding and side II does not appear in the cone of interior directions to C at the point b.

5.3.2 Lemma. Let X be a subset of E_n and K be an open convex cone in E_m. Let α and h be defined on an open set containing X and taking values in E_m and E_k respectively. Suppose that $x_o \in X$ such that $\alpha(x_o) \in C\ell\, K$ and $h(x_o) = 0$. Further suppose that α and h are differentiable at x_o and $\nabla h(x_o)$ has rank k. If $\alpha(x) \in C$, $h(x) = 0$

has no solution in X then $\nabla^t\alpha(x_o)\xi \in K(\alpha(x_o))$, $\nabla^t h(x_o)\xi = 0$ has no solution in $I(X, x_o)$.

Proof: Since $\nabla h(x_o)$ has rank k then $k \leq n$. The case where $k = 0$ corresponds to the absence of equality constraints and the proof is easy. If $k = n$ then the only solution to $\nabla^t h(x_o)\xi = 0$ is $\xi = 0$. Since $I(X, x_o)$ is open then $0 \notin I(X, x_o)$ and the result is immediate. Thus we assume that $0 < k < n$. By the implicit function theorem the following exist:

(i) a partition of $x = (x_1, x_2)$ with $x_1 \in E_{n-k}$, $x_2 \in E_k$ and $\nabla^t h(x_o) = (\nabla_1^t h(x_o), \nabla_2^t h(x_o))$ where $\nabla_2 h(x_o)$ is nonsingular.

(ii) an open set Ω in E_{n-k} containing x_o^1 where $x_o = (x_o^1, x_o^2)$.

(iii) a unique function e on Ω with values in E_k such that $x_o^2 = e(x_o^1)$, e is differentiable on Ω and $h(x^1, e(x^1)) = 0$ for each $x^1 \in \Omega$.

We will prove the lemma by showing that if $\nabla^t\alpha(x_o)\xi \in C(\alpha(x_o))$, $\nabla^t h(h(x_o)\xi = 0$ has a solution in $I(X, x_o)$ then $\alpha(x) \in K$, $h(x) = 0$ has a solution in X. Let ξ be solution to the first system and let $\xi = (\xi_1, \xi_2)$ be the partition of (i) above. Then $\nabla_1^t h(x_o)\xi_1 + \nabla_2^t h(x_o)\xi_2 = 0$. Since $h(x^1, e(x^1)) = 0$ for each $x^1 \in \Omega$ and $x_o^1 \in \Omega$ then $\nabla_1 h(x_o) + \nabla e(x_o)\nabla_2 h(x_o) = 0$ and hence $\nabla_1^t h(x_o)\xi_1 + \nabla_2^t h(x_o)\nabla^t e(x_o^1)\xi_1 = 0$. Since $\nabla_2 h(x_o)$ is nonsingular then the last equation and $\nabla_1^t h(x_o)\xi_1 + \nabla_2^t h(x_o)\xi_2 = 0$ imply that $\xi_2 = \nabla^t e(x_o^1)\xi_1$. By differentiability of e at x_o^1 we have $e(x_o^1 + \delta\xi_1) = e(x_o^1) + \delta\nabla^t e(x_o^1)\xi_1 + \delta \cdot \epsilon_1(\delta)$. Here $\epsilon_1(\delta)$ is a vector valued function of δ such that $\epsilon_1(\delta) \to 0$ as $\delta \to 0$. Thus $e(x_o^1 + \delta\xi_1) = x_o^2 + \delta\xi_2 + \delta \cdot \epsilon_1(\delta)$. Then $(x_o^1 + \delta\xi_1, e(x_o^1 + \delta\xi_1)) = x_o + \delta\xi + \delta \cdot \epsilon_2(\delta)$ where $\epsilon_2(\delta) = (0, \epsilon_1(\delta))$. Now since Ω is open and $x_o^1 \in \Omega$ then for δ sufficiently small we have $x_o^1 + \delta\xi_1 \in \Omega$ and hence $h(x_o^1 + \delta\xi_1, e(x_o^1 + \delta\xi_1)) = 0$. Since α is differentiable at x_o we have $\alpha(x_o^1 + \delta\xi_1, e(x_o^1 + \delta\xi_1)) = \alpha(x_o + \delta\xi + \delta\epsilon_2(\delta)) = \alpha(x_o) + \delta\nabla^t\alpha(x_o)\xi + \delta \cdot \epsilon_3(\delta)$ where $\epsilon_3(\delta) \to 0$ as $\delta \to 0$. Since by assumption $\nabla^t\alpha(x_o)\xi \in K(\alpha(x_o))$ we conclude by Lemma 5.3.1 that $\alpha(x_o^1 + \delta\xi_1, e(x_o^1 + \delta\xi^1)) \in K$ for δ sufficiently small. So far we have shown that $h(x_o + \delta\xi + \delta\, \epsilon_2(\delta)) = 0$ and $\alpha(x_o + \delta\xi + \delta \cdot \epsilon_2(\delta)) \in K$ for δ sufficiently small. Finally since $\xi \in I(X, x_o)$ then for $\delta > 0$ small enough we have $x_o + \delta\xi + \delta \cdot \epsilon_2(\delta) \in X$. Therefore it follows that $(x_o^1 + \delta\xi_1, e(x_o^1 + \delta\xi))$ solves the system $\alpha(x) \in K$, $h(x) = 0$, and $x \in X$ for $\delta > 0$ sufficiently small and the proof is complete.

Now we are able to prove the following theorem which generalizes the Fritz John conditions for optimality.

5.3.3 Theorem. Suppose that x_o solves the problem to minimize $f(x)$ subject to $x \in X$, $g(x) \in C\ell\, C$, and $h(x) = 0$. If f and g are differentiable at x_o and h is continuously differentiable in a neighborhood of x_o then there exists a nonzero $(u_o, u, v) \in E_1 \times E_m \times E_k$ such that

i. $-[\nabla f(x_o)u_o + \nabla g(x_o)u + \nabla h(x_o)v] \in I^*(X, x_o)$

ii. $u_o \geq 0$ $u \in C^*$, and

iii. $\langle u, g(x_o)\rangle = 0$

Proof: If $I(X, x_o)$ is empty then $I^*(X, x_o) = E_n$ and the theorem holds trivially. Without loss of generality assume that the rank of $\nabla h(x_o)$ is k since otherwise the theorem holds trivially. Now suppose that x_o solves the above problem, i.e., $x \in X$, $g(x) \in C\ell\, C$ and $h(x) = 0$ imply that $f(x) \geq f(x_o)$. Let $\alpha(x) = (f(x) - f(x_o), g(x))$ and $K = E_- \times C$ where E_- is the set of negative real numbers. Note that $x_o \in X$, $\alpha(x_o) \in C\ell\, K$ and $h(x_o) = 0$. Also note that the system $\alpha(x) \in K$, $h(x) = 0$ has no solution in X. Therefore by Lemma 5.3.2 the system $\nabla^t\alpha(x_o)\xi \in K(\alpha(x_o))$, $\nabla^t h(x_o)\xi = 0$ has no solution in $I(X, x_o)$. By Lemma 5.1.1 we conclude that there is a nonzero $(q, v) \in K^*(\alpha(x_o)) \times E_k$ such that $\langle q, \nabla^t\alpha(\bar{x})\xi\rangle + \langle v, \nabla^t h(\bar{x})\xi\rangle \geq 0$ for each $\xi \in I(X, x_o)$. Therefore $-[q\nabla^t\alpha(\bar{x}) + v\nabla^t h(\bar{x})] \in I^*(X, x_o)$. Noting that $K(\alpha(x_o)) \supset K$ then $K^*(\alpha(x_o)) \subset K^*$ and so $q \in K^*$, i.e., $q = (u_o, u)$ with $u_o \geq 0$ and $u \in C^*$. Finally since both $\alpha(x_o)$ and $-\alpha(x_o)$ belong to $K(\alpha(x_o))$ and $q \in K^*(\alpha(x_o))$ then $\langle u, g(x_o)\rangle = 0$. To summarize there exist a nonzero (u_o, u, v) such that $u_o \geq 0$, $u \in C^*$, $u^t g(\bar{x}) = 0$, and $-[\nabla f(\bar{x})u_o + \nabla g(\bar{x})u + \nabla h(\bar{x})v] \in I^*(X, x_o)$ and the proof is complete.

The conditions above may be viewed as generalized Fritz John conditions for optimality. The following remarks may be helpful for the reader.

1. u_o, u, and v can be viewed as lagrangian multipliers and (iii) can be viewed as complementarity slackness condition. In the special case when C is the nonnegative orthant then the above conditions reduce to the conditions of Theorem 5.2.2.
2. In the hypothesis of the theorem we require convexity of $I(X, x_o)$. This is

considerably weaker hypotehsis than convexity of X. For example, let $X = \{(x,y) : y \geq x^3\}$ then $I(X, x_o) = \{(x,y) : y > 0\}$. Indeed X is not convex whereas $I(X, x_o)$ is.

3. If $x_o \in \text{int } X$ then $I(X, x_o) = E_n$ and $I^*(X,x_o) = 0$. Then condition (i) becomes $\nabla f(\bar{x})u_o + \nabla g(\bar{x})u + \nabla h(\bar{x})v = 0$. This with conditions (ii) and (iii) give a generalized form of the Fritz John conditions. If C is the non-negative orthant in E_m then we precisely get the Fritz John conditions discussed in Section 2.5.2.

4. In condition (i) of Theorem 5.3.3 if we can replace $I(X, x_o)$ by the larger cone $T(X, x_o)$, i.e., if condition (i) reads $-[\nabla f(x_o)u_o + \nabla g(x_o)u + \nabla h(x_o)v] \in T^*(X, x_o)$, then the theorem will be sharper. However, this sharper result does not hold in general. For example consider the problem: minimize $\{f(x) : x \in X, h(x) = 0\}$ where $X = \{(x_1, x_2) : x_1$ and x_2 are rational$\}$, $f(x_1, x_2) = x_2$, and $h(x_1, x_2) = x_2 - \sqrt{2}\, x_1$. It is clear that $X \cap \{x : h(x) = 0\} = \{(0,0)\}$ so the only admissible point is the origin and hence $x_o = (0,0)$ solves the above problem. It is clear that $T(X, x_o) = E_2$ and hence $T^*(X, x_o) = \{(0,0)\}$. Condition (i), however, does not hold for a nonzero (u_o,v). This shows that we cannot strengthen the conclusion of the theorem by replacing $I(X, x_o)$ with $T(X, x_o)$. One should note that in the above example $I(X, x_o)$ is empty and hence the theorem holds trivially.

Now let us discuss the case when X is a convex set. Then we will show that any $x \in \text{int } X$ implies that $x - x_o \in I(X, x_o)$. Let $x \in \text{int } X$. Then there is a ball N with radius about the origin such that $z \in N$ implies that $x + z \in X$. Now consider $x_o + \lambda(x - x_o + z)$ where $z \in N$. We will show that for every $\lambda \in (0,1)$, $x_o + \lambda(x - x_o + z) \in X$ and thus $x - x_o \in I(X, x_o)$. $x_o + \lambda(x - x_o + z) = (1 - \lambda)x_o + \lambda(x + z) \in X$ by convexity of X and since $x_o \in X$ and $x + z \in X$. Now if X is an open convex set then $x - x_o \in I(X, x_o)$ for each $x \in X$ and so it is clear that condition (i) of the above theorem implies that $\langle x - x_o, u_o\nabla f(x_o) + \nabla g(x_o)u + \nabla h(x_o)v\rangle \geq 0$ for each $x \in X$.

The reader may note that u_o in the above theorem is not necessarily positive. One can pose constraint qualifications that will guarantee positivity of u_o which is discussed in the next chapter.

CHAPTER 6

CONSTRAINT QUALIFICATIONS

We have seen from the last chapter that if a point solves a nonlinear programming problem then the Fritz John conditions hold, provided that the candidate point under investigation belongs to the interior of X. We also showed that if the lagrangian multiplier of the objective function is positive then the Fritz John conditions reduce to the Kuhn-Tucker conditions. In this chapter we will develop in detail various conditions which guarantee positivity of the lagrangian multiplier of the objective function. These conditions are known as constraint qualifications since they only involve the constraints. Section 6.1 below is devoted for problems with inequality constraints while Section 6.2 treats both equality and inequality constraints. Finally a constraint qualification is presented in Section 6.3 in order to obtain necessary optimality criteria of the minimum principle type where the requirement $x_o \in \text{int } X$ is relaxed. The emphasis in this chapter is on the various relationships among different constraint qualifications. The reader may refer to [5 ,28] for a more exhaustive discussion of the different constraint qualifications.

6.1. Inequality Constraint Problems

In this section we deal with problems of the inequality type. Specifically we consider the problem P: minimize $f(x)$ subject to $x \in X$ and $g(x) \le 0$. Here we assume that f and $g = (g_1, g_2, \ldots, g_m)$ are real valued functions which are defined on an open set containing X. We will further assume differentiability of f and $g_i (i \in I)$ at the point x_o under investigation, where $I = \{i : g_i(x_o) = 0\}$. We will denote the feasible region by S, i.e., $S = \{x \in X : g(x) \le 0\}$.

Recall that constraint qualifications are conditions which guarantee that if $x_o \in S$ solves the nonlinear program then the Kuhn-Tucker conditions hold, i.e., there must exist nonnegative scalars $u_i (i \in I)$ with $\nabla f(x_o) + \sum_{i \in I} u_i \nabla g_i(x_o) = 0$. Different qualifications of various strengths will be developed. First we need to use the following cones.

i) $T(S,x_o) = \{x : x = \lim_{k \to \infty} \lambda_k (x_k - x_o)$ where $\lambda_k > 0$ and $x_k \in S$ with $x_k \to x_o\}$. $T(S,x_o)$ is the <u>cone of tangents</u> to S at x_o. $T(S,x_o)$ is closed (see Corollary to Theorem 3.4.4).

ii) $A(S,x_0) = \{x : x_0 + \lambda x + \lambda \epsilon(x;\lambda) \in S$ for each $\lambda \in (0,\delta]$ where $\delta > 0$ and $\epsilon(x;\lambda) \to 0$ as $\lambda \to 0\}$. $A(S,x_0)$ is the <u>cone of attainable directions</u> to S at x_0. (See Definition 3.5.1)

iii) $F(S,x_0) = \{x : x_0 + \lambda x \in S$ for each $\lambda \in (0,\delta]$ where $\delta > 0\}$. $F(S,x_0)$ is the <u>cone of feasible directions</u> to S at x_0. $F(S,x_0)$ is not necessarily closed or open. (See section 3.5.5)

iv) $C_0 = \{x : \langle x, \nabla g_i(x_0)\rangle < 0$ for each $i \in I\}$. It is obvious that C_0 is an open convex cone. The reader may note that when $x_0 \in \text{int } X$ and if, starting with x_0, we move a short distance in the direction of $x \in C_0$, then we will stay in the interior of the feasible region S. More precisely, for $x \in C_0$ there is a $\delta > 0$ such that $x_0 + \lambda x \in \text{int } S$ for all $\lambda \in (0,\delta]$.

v) $C = \{x : \langle x, \nabla g_i(x_0)\rangle \leq 0$ for each $i \in I\}$. It is obvious that C is a closed convex cone. The reader may note that a movement in the direction $x \in C$ may or may not lead to feasible points.

vi) $C_p = \{x : \langle x, \nabla g_i(x_0)\rangle \leq 0$ for each $i \in J$; $\langle x, \nabla g_i(x_0)\rangle < 0$ for each $I \sim J\}$ where $J = \{i \in I : g_i$ is pseudo-concave at $x_0\}$. Obviously C_p is neither closed nor open. The reader may note that $g_i(x_0 + \lambda x) \leq 0$ for all i provided that $x \in C_p$ and $\lambda > 0$ is sufficiently small.

In order to develop the relationships and the implications among the various constraint qualifications which we will pose later, the relationships between the above cones are first established in Lemma 6.1.1 and Theorem 6.1.2.

6.1.1 <u>Lemma</u>. Let $x_0 \in \text{int } X$ and let g_i be differentiable at x_0 for $i \in I$ and g_i be continuous at x_0 for $i \notin I$, where $I = \{i : g_i(x_0) = 0\}$. Then $C^* \subset T^*(S,x_0)$.

<u>Proof</u>: Let $G_i = \{x : g_i(x) \leq 0\}$. Then $S = X \cap \bigcap_{i=1}^{m} G_i$ and hence $T(S,x_0) = T(X \cap \bigcap_{i=1}^{m} G_i, x_0) \subset T(X,x_0) \cap \bigcap_{i=1}^{m} T(G_i,x_0)$. Noting that for $i \notin I, g_i(x_0) < 0$ then by continuity of g_i at x_0 it follows that $x_0 \in \text{int } G_i$ and hence $T(G_i,x_0) = E_n$ for each $i \notin I$. Also since $x_0 \in \text{int } X$ then $T(X,x_0) = E_n$ and so we get $T(S,x_0) \subset \bigcap_{i \in I} T(G_i,x_0)$. Therefore $T^*(S,x_0) \supset (\bigcap_{i \in I} T(G_i,x_0))^* \supset \sum_{i \in I} T^*(G_i,x_0)$ by Theorem

3.1.10.

We will first now that $T^*(G_i,x_o) \supset C_i^*$ where $C_i = \{x : \langle x,\nabla g_i(x_o)\rangle \le 0\}$ for $i \in I$. Let $x \in T(G_i,x_o)$, i.e., $x = \lim_{k\to\infty} \lambda_k(x_k - x_o)$ where $\lambda_k > 0$, $g_i(x_k) \le 0$ and $x_k \to x_o$. But $g_i(x_k) = g_i(x_o) + \langle x_k - x_o,\nabla g_i(x_o)\rangle + \|x_k - x_o\|\cdot\epsilon(x_k - x_o)$. Multiplying by $\lambda_k > 0$ and letting $k \to \infty$ we get $\langle x,\nabla g_i(x_o)\rangle \le 0$. Therefore $x \in C_i$ and hence $T(G_i,x_o) \subset C_i$. Therefore $T^*(G_i,x_o) \supset C_i^*$ and hence $T^*(S,x_o) \supset \sum_{i\in I} C_i^*$. But since the C_i's are polyhedral cones and since $C = \bigcap_{i\in I} C_i$ it follows by Theorem 3.2.6 that $C^* = \sum_{i\in I} C_i^*$ and hence $T^*(S,x_o) \supset C^*$. This completes the proof.

<u>Corollary</u>. $C \supset T^{**}(S,x_o)$.

Since $C_i^* = \{\lambda\nabla g_i(x_o) : \lambda \ge 0\}$ it is clear from the above lemma that any point of the form $\sum_{i\in I} u_i\nabla g_i(x_o)$ belongs to $T^*(S,x_o)$ as long as $u_i \ge 0$ for each $i \in I$. It is interesting to note that the converse of the above lemma does not hold in general. In fact the condition $C \subset T^{**}(S,x_o)$ is the constraint qualification of Guignard, which is the weakest possible qualification for inequality problems.

<u>6.1.2 Theorem</u>. $C_o \subset C_p \subset C$ and $F(S,x_o) \subset A(S,x_o) \begin{matrix} \subset A^{**}(S,x_o) \\ \subset T(S,x_o) \end{matrix} \subset T^{**}(S,x_o)$. Furthermore if $x_o \in \text{int } X$ then $C_o \subset C_p \subset F(S,x_o) \subset A(S,x_o) \begin{matrix} \subset A^{**}(S,x_o) \\ \subset T(S,x_o) \end{matrix} \subset T^{**}(S,x_o) \subset C$.

<u>Proof</u>: It is clear that $C_o \subset C_p \subset C$. Also the inclusions $F(S,x_o) \subset A(S,x_o) \begin{matrix} \subset A^{**}(S,x_o) \\ \subset T(S,x_o) \end{matrix} \subset T^{**}(S,x_o)$ follow from the definitions of these cones. To complete the proof we need to show that if $x_o \in \text{int } X$ then $C_p \subset F(S,x_o)$ and $T^{**}(S,x_o) \subset C$. The latter inclusion follows immediately from Lemma 6.1.1. Now we show that $C_p \subset F(S,x_o)$ if $x_o \in \text{int } X$. Now let $x \in C_p$, i.e., $\langle x,\nabla g_i(x_o)\rangle \le 0$ for $i \in J$ and $\langle x,\nabla g_i(x_o)\rangle < 0$ for $i \in I \sim J$ where $J = \{i: g_i$ is pseudo-concave at $x_o\}$. Consider $x_o + \lambda x$ for $\lambda > 0$. Since $x_o \in \text{int } X$ then $x_o + \lambda x \in X$ for λ sufficiently small. By continuity of $g_i (i \notin I)$ and since $g_i(x_o) < 0$ it is clear that $g_i(x_o + \lambda x) < 0$ for λ sufficiently small. For the pseudo-concave binding constraints, $\langle x,\nabla g_i(x_o)\rangle \le 0$ implies that $g_i(x_o + \lambda x) \le g_i(x_o) = 0$ for $\lambda > 0$. Finally for $i \in I \sim J$, $\langle x,\nabla g_i(x_o)\rangle < 0$ implies that $g_i(x_o + \lambda x) < 0$ for λ sufficiently small. In other words $x_o + \lambda x \in S$ for $\lambda > 0$ and sufficiently small, i.e., $x \in F(S,x_o)$ and

the proof is complete.

Before proceeding any further it may be helpful to show that in general we can find no implications among $A^{**}(S,x_o)$ and $T(S,x_o)$. We first give an example where $A^{**}(S,x_o) \supset T(S,x_o)$ but $A^{**}(S,x_o) \neq T(S,x_o)$. Here we let $X = E_n$, $g_1(x_1,x_2) = x_1x_2$, $g_2(x_1,x_2) = -x_1$ and $g_3(x_1,x_2) = -x_2$. In other words, $S = \{(x_1,x_2) : x_1 = 0, x_2 \geq 0\} \cup \{(x_1,x_2) : x_2 = 0, x_1 \geq 0\}$. Now consider the origin. $T(S,(0,0)) = S$ and $A(S,(0,0)) = S$. However $A^{**}(S,(0,0)) = \{(x_1,x_2) : x_1 \geq 0, x_2 \geq 0\}$. Clearly $T(S,(0,0)) \subset A^{**}(S,(0,0))$ but they are not equal. We will now give another example where $A^{**}(S,x_o) \subset T(S,x_o)$ but are unequal. Consider the following due to Abadie [1].

$$\begin{aligned} s(x_1) &= x_1^4 \sin \frac{1}{x_1} & x_1 &\neq 0 \\ &= 0 & x_1 &= 0 \end{aligned} \qquad \text{and}$$

$$\begin{aligned} c(x_1) &= x_1^4 \cos \frac{1}{x_1} & x_1 &\neq 0 \\ &= 0 & x_1 &= 0 \end{aligned}$$

The above two functions are continuously differentiable. The functions and their derivatives vanish at the point $x_1 = 0$. Now consider the following constraint functions.

$$\begin{aligned} g_1(x_1,x_2) &= x_2 - x_1^2 - s(x_1) \\ g_2(x_1,x_2) &= -x_2 + x_1^2 + c(x_1) \\ g_3(x_1,x_2) &= x_1^2 - 1 \end{aligned}$$

Here we take X as E_n. The reader can verify that $A(S,(0,0)) = A^{**}(S,(0,0)) = \{(0,0)\}$ whereas $T(S,(0,0)) = \{(x_1,x_2) : x_2 = 0\}$.

Various Constraint Qualifications

We will now introduce some of the constraint qualifications which insures that if a point $x_o \in X$ solves the nonlinear programming problem minimize $\{f(x) : x \in X, g_i(x) \leq 0 \quad i = 1,2,\ldots,m\}$ then the Kuhn-Tucker conditions must hold, i.e., there must exist a nonnegative vector $u = (u_1,u_2,\ldots,u_m)$ such that:

$$\text{i.} \quad \nabla f(x_o) + \sum_{i=1}^{m} u_i \nabla g_i(x_o) = 0$$

$$\text{ii.} \quad u_i g_i(x_o) = 0 \quad i = 1,2,\ldots,m.$$

1. Cottle Constraint Qualification [8].

$x_o \in \text{int } X$ and $C_o \neq \emptyset$ or $x_o \in \text{int } X$ and $C \subset C\ell\, C_o$.

In its original form Cottle's constraint qualification requires that the system $\sum_{i \in I} u_i \nabla g_i(x_o) = 0$ has no nonnegative solution which is different from the zero vector. This is the same as stating that C_o is not empty, i.e., there is a vector x such that $\langle x, \nabla g_i(x_o)\rangle < 0$ for each $i \in I$. For suppose by contradiction there is no such vector x, then by Theorem 5.1.2 the system $\sum_{i \in I} u_i \nabla g_i(x_o) = 0$ has a solution which is nonnegative and not zero, violating the constraint qualification. Also note that the statement $C_o \neq \emptyset$ implies that $C\ell\, C_o = \bigcap_{i \in I} \{x : \langle x, \nabla g_i(x_o)\rangle \leq 0\} = C$. Hence the relationship $C \subset C\ell\, C_o$ is another form of the constraint qualification of Cottle.

2. Arrow-Hurwicz-Uzawa - I Constraint Qualification [2]

$x_o \in \text{int } X$ and $C_p \neq \emptyset$ or $x_o \in \text{int } X$ and $C \subset C\ell\, C_p$.

This constraint qualification was originally given by Arrow-Hurwicz-Uzawa where the set J corresponds to indices of concave functions. This was relaxed later [25] to pseudo-concave functions at the points x_o. The qualification requires the system $\langle x, \nabla g_i(x_o)\rangle \leq 0$ for $i \in J$ and $\langle x, \nabla g_i(x_o)\rangle < 0$ for $I \sim J$ to have a solution. It can be shown that the condition $C_p \neq \emptyset$ is equivalent to $C \subset C\ell\, C_p$.

3. Zangwill Constraint Qualification [36]

$C \subset C\ell\, F(S, x_o)$.

4. Kuhn-Tucker Constraint Qualification [23]

$C \subset C\ell\, A(S, x_o)$.

5. Arrow-Hurwicz-Uzawa - II Constraint Qualification [2]

$C \subset A^{**}(S, x_o)$.

6. Abadie Constraint Qualification [1]

$C \subset T(S, x_o)$.

7. Guignard Constraint Qualification [19]

$C \subset T^{**}(S, x_o)$.

From the above definitions and by Theorem 6.1.2 it is clear that the relationships among the above constraint qualifications are as given in Figure 6.1. Clearly the weakest among these is Guignard constraint qualification whereas the strongest is Cottle's constraint qualification. Perhaps the more important fact to observe from the above theorem and the above qualifications is the closeness of the above qualifications in the sense that if one of them is satisfied then it is likely that the others are also satisfied for most porblems. This follows from the fact that C_o = int C and hence all the cones given above lie indeed between a set and its interior.

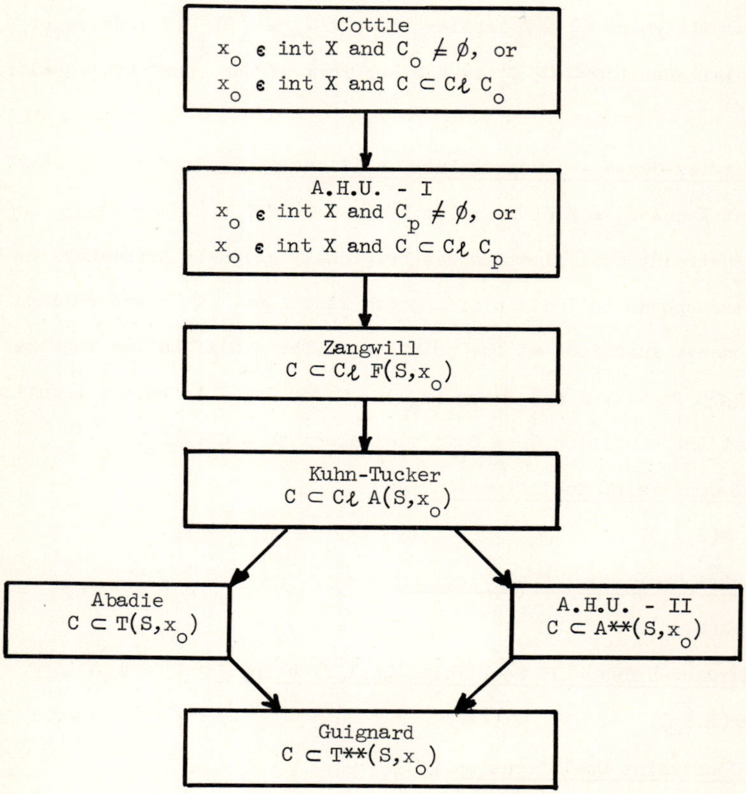

Figure 6.1

Relations Among Different Constraint Qualifications For Inequality Constraint Problems

At this stage it may be helpful to give an example where none of the above constraint qualifications hold. Consider the problem of minimizing $f(x_1,x_2) = x_1$ subject to $g_1(x_1,x_2) = x_2 \leq 0$ and $g_2(x_1,x_2) = -(x_1 - 1)^3 + x_2 \leq 0$. Clearly the optimal point x_o is $(1,0)$. Note that $T(S,x_o) = \{(x_1,x_2) : x_1 \geq 0, x_2 = 0\}$ and $T^*(S,x_o) = \{(x_1,x_2) : x_1 \leq 0\}$. On the other hand $\nabla g_1(1,0) = (0, -1)$ and $\nabla g_2(1,0) = (0,1)$ and hence $C^* = C_1^* + C_2^* = \{(x_1,x_2) : x_1 = 0\}$. Therefore $T^*(S,x_o) \neq C^*$ and Guignard qualification does not hold. This also implies that none of the other qualifications will hold. Obviously the Kuhn-Tucker conditions do not hold even though the point $(1,0)$ is optimal.

So far we have posed different constraint qualifications and we showed their significance and interrelationships. Theorem 6.1.4 below shows that these qualification guarantee that the Kuhn-Tucker conditions indeed hold at optimality. In view of Figure 6.1 it suffices to assume that Guignard qualification holds. First the following lemma is needed.

6.1.3 <u>Lemma</u>. Let x_o solve the problem minimize $f(x)$ subject to $x \in S$. Suppose that f is differentiable at x_o. Then $-\nabla f(x_o) \in T^*(S,x_o)$.

<u>Proof</u>: Let $x \in T(S,x_o)$. We need to show that $\langle x,\nabla f(x_o)\rangle \geq 0$. But $x \in T(S,x_o)$ implies that $x = \lim_{k\to\infty} \lambda_n(x_k - x_o)$ with $\lambda_k > 0$, $x_k \in S$ and $x_k \to x_o$. By differentiability of f at x_o we get

$$f(x_k) = f(x_o) + \langle x_k - x_o,\nabla f(x_o)\rangle + \|x_k - x_o\|\cdot\epsilon(x_k - x_o)$$

where $\epsilon(x_k - x_o) \to 0$ as $n\to\infty$. Since x_o solves the problem and $x_k \in S$ then $f(x_k) \geq f(x_o)$. This shows that $\langle x_k - x_o,\nabla f(x_o)\rangle + \|x_k - x_o\|\cdot\epsilon(x_k - x_o) \geq 0$. Multiplying by $\lambda_k > 0$ and letting $k \to \infty$ we get $\langle x,\nabla f(x_o)\rangle \geq 0$ and the result is at hand.

<u>6.1.4 Theorem</u>. (Kuhn-Tucker Conditions). Suppose that x_o solves the problem: minimize $\{f(x) : x \in X, g_i(x) \leq 0 \; i = 1,2,\dots,m\}$. Let $I = \{i : g_i(x_o) = 0\}$ and suppose that f and $g_i(i \in I)$ are differentiable at x_o and $g_i(i \notin I)$ are continuous at x_o. Let $S = \{x \in X : g_i(x) \leq 0 \quad i = 1,2,\dots,m\}$. Suppose that Guignard constraint qualification $C \subset T^{**}(S,x_o)$ holds. Then there exists a vector $(u_1,u_2,\dots,u_m)$ such

that:

i. $\nabla f(x_o) + \sum_{i=1}^{m} u_i \nabla g_i(x_o) = 0$

ii. $u_i g_i(x_o) = 0 \qquad i = 1,2,\ldots,m$

iii. $u_i \geq 0 \qquad i = 1,2,\ldots,m$

Proof: Since x_o solves the problem then by Lemma 6.1.3, $-\nabla f(x_o) \in T^*(S,x_o)$. Since $C \subset T^{**}(S,x_o)$ then $T^*(S,x_o) \subset C^*$ and so $-\nabla f(x_o) \in C^* = \sum_{i \in I} C_i^*$. But since $C_i^* = \{\lambda \nabla g_i(x_o) : \lambda \geq 0\}$ then $-\nabla f(x_o) = \sum_{i \in I} u_i \nabla g_i(x_o)$ where $u_i \geq 0$ for each $i \in I$. Letting $u_i = 0$ for $i \notin I$ the result follows.

From Figure 6.1 and the above theorem it is immediate that any of the constraint qualifications given earlier will insure that the Kuhn-Tucker conditions hold provided that the point is an optimal solution.

The reader may note that the assumption $x_o \in \text{int } X$ is not required in Theorem 6.1.4 above. However, without this assumption it is very unlikely that the qualification $T^*(S,x_o) \subset C^*$ will hold. A more suitable assumption for the case when x_o is not necessarily in the interior of X is $T^*(S,x_o) \subset C^* + T^*(X,x_o)$ (note that $T^*(S,x_o) \supset C^* + T^*(X,x_o)$ always hold). In this case we will not get the Kuhn-Tucker conditions but rather conditions of the form $\nabla f(x_o) + \sum_{i=1}^{m} u_i \nabla g_i(x_o) \in T^*(X,x_o)$. Clearly if $x_o \in \text{int } x_o$ then $T^*(X,x_o) = \{0\}$ and we get the Kuhn-Tucker conditions. To illustrate, consider the example of Figure 6.2. Here we have the nonfunctionally specified set X as well as the constraint function $g(x_1,x_2) = -x_1 + x_2^2$. The objective is to minimize $2x_1 + 3x_2$. In other words we want to solve the problem : minimize $\{2x_1 + 3x_2 : (x_1,x_2) \in X, - x_1 + x_2^2 \leq 0\}$. It is clear that the optimal point is the origin (0,0). It is clear that $T^*(S,x_o) = \{(x_1,x_2) : x_1 \leq 0, x_2 \leq 0\}$. $\nabla g(0,0) = (-1,0)$ and so $C^* = \{\lambda(-1,0) : \lambda \geq 0\}$. Clearly then $T^*(S,x_o) \not\subset C^*$, i.e., the constraint qualification does not hold. Actually we should note that $(0,0) \notin \text{int } X$. The Kuhn-Tucker conditions do not hold because $\nabla f(x_o) + u\nabla g(x_o) = (2,3) + u(-1,0) = (0,0)$ has no solution. As a matter of fact the Fritz John conditions do not hold either since the system $u_o(2,3) + u(-1,0) = (0,0)$ has no nonzero solution.

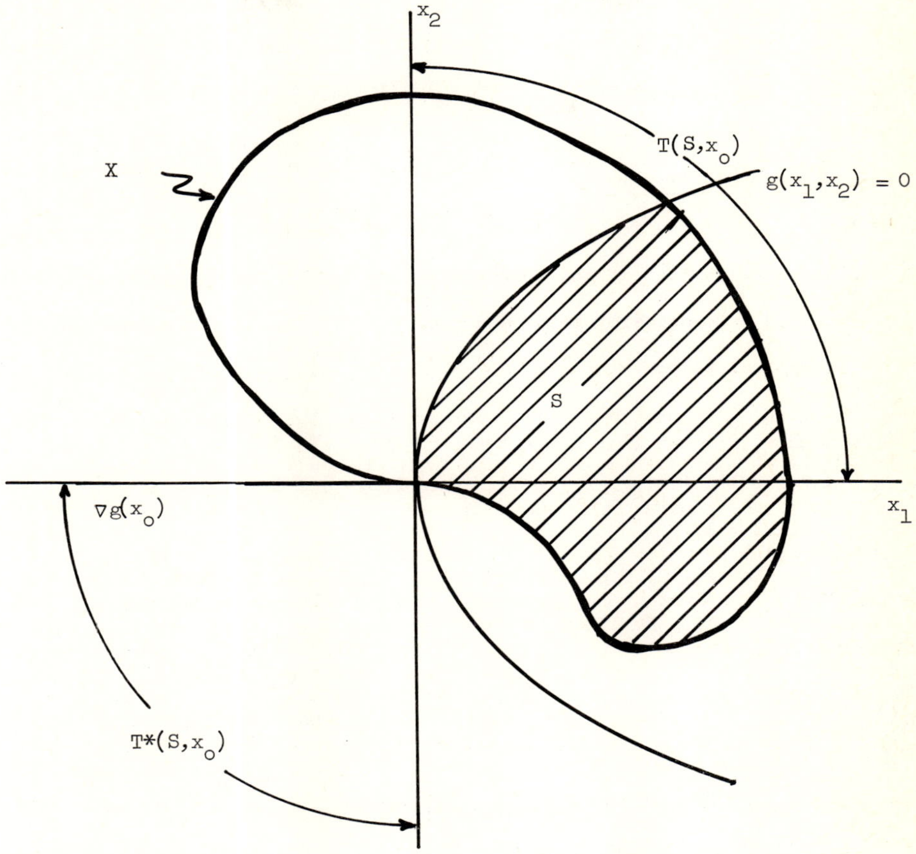

Figure 6.2

An Example Where $x_o \notin$ int X

On the other hand suppose that we consider the constraint qualification $T^*(S,x_o) \subset T^*(X,x_o) + C^*$. From Figure 6.2 it is clear that $T^*(S,x_o) = \{(x_1,x_2) : x_1 \le 0,\ x_2 \le 0\}$ whereas $T^{**}(X,x_o) = \{(x_1,x_2) : x_1 = 0,\ x_2 \le 0\}$. Since $C^* = \{(x_1,x_2) : x_2 = 0,\ x_1 \le 0\}$ it is clear that $T^*(S,x_o) = T^*(X,x_o) + C^*$, i.e., the constraint qualification holds. Under this we know that there must exist a $u \ge 0$ such that $-[\nabla f(x_o) + u\nabla g(x_o)] \in T^*(S,x_o)$. It is obvious that each u in the interval $[0,2]$ will serve the purpose.

Case of Linear Constraints

Now we would like to discuss the important special case of linear constraints. We will show below that if the constraints are linear then Zangwill constraint qualification automatically holds and therefore the Kuhn-Tucker conditions must hold at every optimal point.

6.1.5 Lemma. Let $S = \{x : g_i(x) \le 0\ i = 1,2,\ldots,m\}$ where $g_i(x) = \langle a_i,x\rangle - b_i$. Let $x_o \in S$ and let $I = \{i : g_i(x_o) = 0\}$. Then $F(S,x_o) = C$ where $C = \{x : \langle x,\nabla g_i(x_o)\rangle \le 0 \text{ for each } i \in I\}$.

Proof: Let $x \in C$, i.e., $\langle x,\nabla g_i(x_o)\rangle = \langle x,a_i\rangle \le 0$ for each $i \in I$. Now consider $x_o + \lambda x$ where $\lambda > 0$. By linearity of g_i we have $g_i(x_o + \lambda x) = g_i(x_o) + \lambda\langle x,\nabla g_i(x_o)\rangle = \lambda\langle x,a_i\rangle \le 0$ for each $\lambda > 0$ and each $i \in I$. Now for $i \notin I, g_i(x_o) < 0$ and $g_i(x_o + \lambda x) \le 0$ for λ sufficiently small. This means that $g_i(x_o + \lambda x) \le 0$ for $i = 1,2,\ldots,m$ for $\lambda > 0$ sufficiently small. In other words $x \in F(S,x_o)$. Therefore $C \subset F(S,x_o)$ and since $C \supset F(S,x_o)$ the proof is complete.

From the above we know that if a point solves a program where the constraints are linear (the objective function may be linear or nonlinear) then the Kuhn-Tucker conditions will always hold. In particular if we have a linear problem of the form minimize $\langle c,x\rangle$ subject to $Ax \le b$ and $x \ge 0$ then the Kuhn-Tucker conditions read as follows. There exist a nonnegative vector u and v such that:

$$c + A^t u - v = 0$$

$$\langle u,Ax - b\rangle = 0$$

$$\langle v,x\rangle = 0$$

The vector v corresponding to the nonnegativity constraints may be deleted and the conditions become $c + A^t u \ge 0$, $\langle u,Ax - b\rangle = 0$, and $\langle c + A^t u,x\rangle = 0$.

Other Weaker Constraint Qualifications

The reader may note that the above constraint qualifications are put in terms of set inclusions. There are other types of constraints which depend upon the nature of the binding constraints, e.g., convexity or concavity. We will give below some of these constraint qualifications.

1. Slater Constraint Qualification [32]

$x_o \in \text{int } X$, $g_i (i \in I)$ are pseudo-convex at x_o, and there exists an $x \in X$ with $g_i(x) < 0$ for each $i \in I$.

2. Reverse Constraint Qualification [25]

$x_o \in \text{int } X$, and $g_i (i \in I)$ are pseudo-concave at x_o.

3. Independence Constraint Qualification [3]

$x_o \in \text{int } X$, and $\nabla g_i(x_o)$ for $i \in I$ are linearly independent.

4. Karlin Constraint Qualification [22]

$x_o \in \text{int } X$, $g_i (i \in I)$ are convex at x_o, and there is no nonzero nonnegative vector u_I with components $u_i (i \in I)$ such that $\sum_{i \in I} u_i g_i(x) \geq 0$ for each $x \in X$.

The following theorem shows the implications among these constraint qualifications and it also shows the relationships between these constraint qualifications and the constraint qualifications we posed earlier in the section.

6.1.6 Theorem.

i. Slater constraint qualification $\rightarrow$ Cottle constraint qualification.

ii. Reverse constraint qualification $\rightarrow$ Zangwill constraint qualification.

iii. Independence constraint qualification $\rightarrow$ Cottle constraint qualification.

iv. Karlin constraint qualification $\rightarrow$ Cottle constraint qualification.

Proof:

i. Suppose that Slater constraint qualification holds at x_o, i.e., there is an $x_o \in \text{int } X$, $g_i (i \in I)$ are pseudo-convex at x_o, and there exists an $x \in X$ with $g_i(x) < 0$ for each $i \in I$. By contradiction suppose that Cottle constraint qualification does not hold. Then there must exist a nonzero solution to the system $\sum_{i \in I} u_i \nabla g_i(x_o) = 0$ where $u_i \geq 0$ for each $i \in I$.

It may be noted that $\langle x - x_o, \nabla g_i(x_o)\rangle < 0$ for all $i \in I$ because if not then $\langle x - x_o, \nabla g_i(x_o)\rangle \geq 0$ for some $i \in I$ and by pseudo-convexity of g_i at x_o it follows that $g_i(x) \geq g_i(x_o) = 0$ which is impossible since $g_i(x) < 0$. Since $u_i \geq 0$ for each $i \in I$ and not all the u_i's are zero then $\sum_{i \in I} u_i \langle x - x_o, \nabla g_i(x_o)\rangle < 0$, which contradicts the assumption that $\sum_{i \in I} u_i \nabla g_i(x_o) = 0$.

ii. Suppose that the reverse constraint qualification is satisfied at x_o, i.e., $x_o \in \text{int } X$ and $g_i (i \in I)$ are pseudo-concave at x_o. Let $x \in C$, i.e., $\langle x, \nabla g_i(x_o)\rangle \leq 0$ for every $i \in I$. We show that $x \in F(S, x_o)$, i.e., $x_o + \lambda x \in S$ for $\lambda > 0$ and sufficiently small. Since $x_o \in \text{int } X$ then $x_o + \lambda x \in X$ for λ sufficiently small. By continuity of $g_i (i \notin I)$ and since $g_i(x_o) < 0$ then $g_i(x_o + \lambda x) \leq 0$ for λ sufficiently small and $i \notin I$. Finally by pseudo-concavity of $g_i (i \in I)$ at x_o, $\langle x_o + \lambda x - x_o, \nabla g_i(x_o)\rangle = \lambda \langle x, \nabla g_i(x_o)\rangle \leq 0$ implies that $g_i(x_o + \lambda x) \leq 0$ for $i \in I$ and $\lambda > 0$. This shows that $x_o + \lambda x \in S$ for $\lambda > 0$ and sufficiently small. Hence Zangwill constraint qualification holds.

iii. Obvious.

iv. We will show that Karlin qualification implies Cottle constraint qualification by showing that if Cottle qualification does not hold then Karlin qualification will not hold. Suppose that the system $\sum_{i \in I} u_i \nabla g_i(x_o) = 0$ has a nonzero nonnegative solution. By convexity of g_i at x_o for $i \in I$ it follows that $g_i(x) \geq g_i(x_o) + \langle x - x_o, \nabla g_i(x_o)\rangle = \langle x - x_o, \nabla g_i(x_o)\rangle$ for every $x \in X$. Multiplying by $u_i \geq 0$ and adding over $i \in I$ we get $\sum_{i \in I} u_i g_i(x) \geq \langle x - x_o, \sum_{i \in I} u_i \nabla g_i(x_o)\rangle = 0$. This shows that the system $\sum_{i \in I} u_i g_i(x) \geq 0$ for every $x \in X$ has a nonzero nonnegative solution and the proof is complete.

Putting together the relationships between the above constraint qualifications and those of Figure 6.1, Figure 6.3 below summarizes the relationship between the different constraint qualifications.

6.2. Equality and Inequality Constraint Problems

In this section we consider a problem of the form: minimize $f(x)$ subject to $x \in X$, $g(x) \leq 0$, and $h(x) = 0$. Here $g = (g_1, g_2, \ldots, g_m)$ denotes the inequality constraints and $h = (h_1, h_2, \ldots, h_k)$ denotes the equality constraints. X denotes an

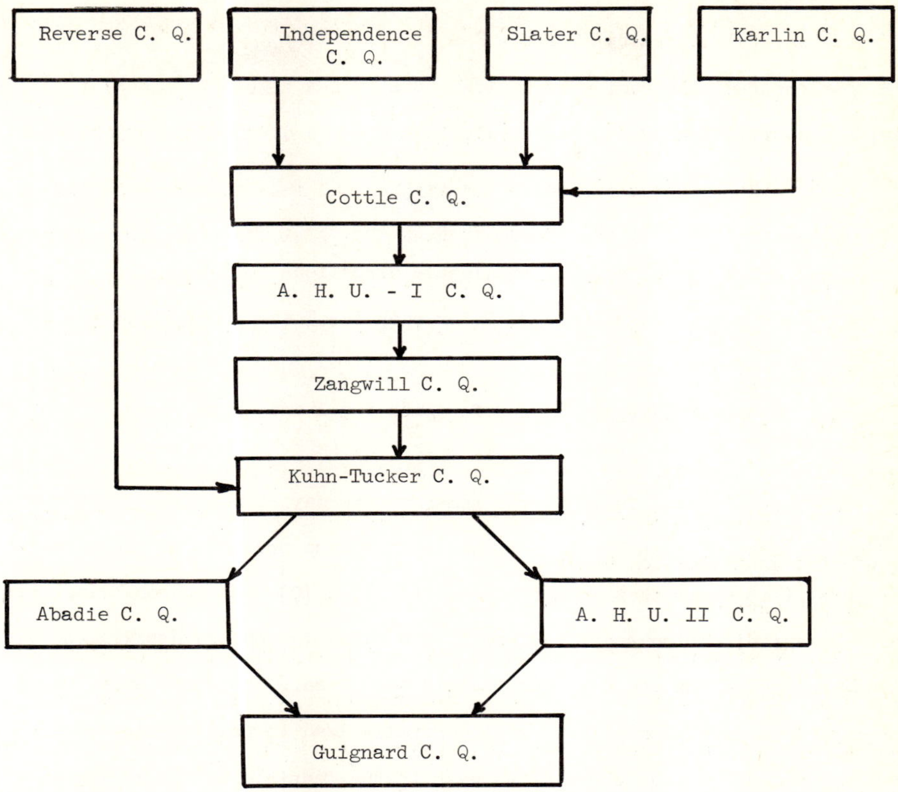

Figure 6.3

Relationship Between Various Constraint Qualifications

For Inequality Constraint Problems

arbitrary set. Corresponding to the point x_0 under consideration we let I be the set of binding indices, i.e., $I = \{i : g_i(x_0) = 0\}$. For convenience we denote the set of equality indices by E, i.e., $E = \{1,2,\ldots,k\}$.

Let the feasible region be S, i.e., $S = \{x \in X : g(x) \leq 0,\ h(x) = 0\}$. Let $x_0 \in S$. We will make use of the following cones in discussing various constraint qualifications.

i. $T(S,x_0)$, the cone of tangents to the set S at x_0.

ii. $A(S,x_0)$, the cone of attainable directions to S at x_0.

iii. $C_0 = \{x : \langle x, \nabla g_i(x_0)\rangle < 0$ for each $i \in I$ and $\langle x, \nabla h_i(x_0)\rangle = 0$ for each $i \in E\}$.

iv. $C = \{x : \langle x, \nabla g_i(x_0)\rangle \leq 0$ for each $i \in I$ and $\langle x, \nabla h_i(x_0)\rangle = 0$ for each $i \in E\}$.

Before proceeding further some remarks here may be helpful. First we do not define the cone of feasible directions. Due to the presence of the equality constraints it is clear that in most cases $F(S,x_0) = \{0\}$, the reason being that if one moves in any direction starting from x_0 the equality constraints will be violated, unless of course the equlaity constraints are linear. For this same reason, a movement in a direction x in C_0 will generally lead to points which are not feasible. However, we can keep as close as we wish to the feasible set by moving in a direction in C_0. As we have done before we can relax C_0 to the cone C_p where the pseudo-convex binding constraints are distinguished from the nonpseudo-convex binding constraints. More precisely one may consider $C_p = \{x : \langle x, \nabla g_i(x_0)\rangle < 0$ for $i \in J$, $\langle x, \nabla g_i(x_0)\rangle \leq 0$ for $i \in I \sim J$, and $\langle x, \nabla h_i(x_0)\rangle = 0$ for $i \in E\}$. However we choose not to do so because this generalization follows in an obvious manner as in Section 6.1.

In showing the relationships between some of the constraint qualifications we need the result $C^* \subset T^*(S,x_0)$. This result was proved by Lemma 6.1.1 for the case when $S = \{x : g(x) \leq 0\}$ and $C = \{x : \langle x, \nabla g_i(x_0)\rangle \leq 0$ for each $i \in I\}$. The following lemma states that indeed $C^* \subset T^*(S,x_0)$ holds in the presence of equality constraints, i.e., when $S = \{x \in X : g(x) \leq 0,\ h(x) = 0\}$ and $C = \{x : \langle x, \nabla g_i(x_0)\rangle \leq 0$ for each $i \in I$ and $\langle x, \nabla h_i(x_0)\rangle = 0$ for each $i \in E\}$. The proof is similar to that of Lemma 6.1.1 and is left for the reader as an exercise.

6.2.1 Lemma. Let $x_o \in \text{int } X$, $g_i (i \in I)$ be differentiable at x_o, $g_i (i \notin I)$ be continuous at x_o, and $h_i (i \in E)$ be differentiable at x_o. Then $C^* \subset T^*(S,x_o)$.

6.2.2 Lemma. $A(S,x_o) \begin{matrix} \subset A^{**}(S,x_o) \\ \subset T(S,x_o) \end{matrix} \subset T^{**}(S,x_o) \subset C$. Furthermore if $\nabla h_i(x_o)$ for $i \in E$ are linearly independent then $C_o \subset A(S,x_o)$.

Proof: $A(S,x_o) \begin{matrix} \subset A^{**}(S,x_o) \\ \subset T(S,x_o) \end{matrix} \subset T^{**}(S,x_o)$ follow trivially from the definitions of these cones. By Lemma 6.2.1 $C^* \subset T^*(S,x_o)$ and so $C^{**} \supset T^{**}(S,x_o)$. Noting that $C^{**} = C$ since the latter is a closed convex cone then $T^{**}(S,x_o) \subset C$. To complete the proof we need to show that under the independence assumption of $\nabla h_i(x_o)$ for $i \in E$, $C_o \subset A(S,x_o)$. So let $x \in C_o$, i.e., $\langle x, \nabla g_i(x_o)\rangle < 0$ for $i \in I$ and $\langle x, \nabla h_i(x_o)\rangle = 0$ for $i \in E$. As in the proof of Theorem 5.2.1 we can construct a feasible arc α with $\alpha(0) = x_o$ and $\frac{d\,\alpha(\theta)}{d\,\theta} = x$ and hence $x \in A(S,x_o)$. This completes the proof.

We will now present some of the constraint qualifications corresponding to a feasible point $x_o \in S$. Some of these qualifications are in the form of set inclusion and some will be given via suitable convexity requirements. These qualifications extend the corresponding qualifications for inequality problems.

1. Cottle Constraint Qualification

   $x_o \in \text{int } X$, $C_o \neq \emptyset$, and $\nabla h_i(x_o)$ for $i \in E$ are linearly independent. This can be equivalently stated as: $x_o \in \text{int } X$, $C\ell\, C_o = C$, and $\nabla h_i(x_o)$ for $i \in E$ are linearly independent.

2. Kuhn-Tucker Constraint Qualification

   $C \subset C\ell\, A(S,x_o)$.

3. Arrow-Hurwicz-Uzawa Constraint Qualification

   $C \subset A^{**}(S,x_o)$.

4. Abadie Constraint Qualification

   $C \subset T(S,x_o)$.

5. Guignard Constraint Qualification

   $C \subset T^{**}(S,x_o)$.

6. Slater Constraint Qualification

   $x_o \in \text{int } X$, $g_i (i \in I)$ are pseudo-convex at x_o, $h_i (i \in E)$ are linear, and there exists an $x \in X$ with $g_i(x) < 0$ for each $i \in I$ and $h_i(x) = 0$ for each $i \in E$.

7. Reverse Constraint Qualification

$x_o \in \text{int } X$, $g_i (i \in I)$ are pseudo-concave at x_o and $h_i (i \in E)$ are pseudo-linear (i.e., pseudo-convex and pseudo-concave) at x_o.

8. Independence Constraint Qualification

$x_o \in \text{int } X$ and $\nabla g_i(x_o)$ for $i \in I$ and $\nabla h_i(x_o)$ for $i \in E$ are linearly independent.

9. Karlin Constraint Qualification

$x_o \in \text{int } X$, $g_i (i \in I)$ are convex at x_o, $h_i (i \in E)$ are linear and there exist no nonzero vector (u_I, v) with $\sum_{i \in I} u_i g_i(x) + \sum_{i \in E} v_i h_i(x) \geq 0$ for each $x \in X$, where u_I is a nonnegative vector of components u_i for $i \in I$ and $v = (v_1, v_2, \ldots, v_k)$.

The following theorem shows the relationships among the above qualifications. It will be clear that Guignard constraint qualification is the weakest among them. We will then show via Theorem 6.2.4 below that under Guignard qualification (and hence under any of the other qualifications) the Kuhn-Tucker conditions hold, provided of course that the point under investigation is an optimal solution.

6.2.3 Theorem.

i. Cottle $\rightarrow$ Kuhn-Tucker $\rightarrow$ Arrow-Hurwicz-Uzawa $\rightarrow$ Abadie $\rightarrow$ Guignard

ii. Slater $\rightarrow$ Cottle

iii. Independence $\rightarrow$ Cottle

iv. Karlin $\rightarrow$ Cottle

v. Reverse $\rightarrow$ Kuhn-Tucker

Proof: i follows trivially from Lemma 6.2.2 and the definitions of the corresponding constraint qualifications. To prove ii suppose that Slater constraint qualification holds. Let $\hat{x} \in X$ be such that $g_i(\hat{x}) < 0$ for each $i \in I$ and $h_i(\hat{x}) = 0$ for $i \in E$. By pseudo-convexity of $g_i (i \in I)$ at x_o and since $g_i(\hat{x}) < g_i(x_o) = 0$ then $\langle \hat{x} - x_o, \nabla g_i(x_o) \rangle < 0$. By linearity of h_i under Slater qualification we have $h_i(\hat{x}) = h_i(x_o) + \langle \hat{x} - x_o, \nabla h_i(x_o) \rangle$. But $h_i(\hat{x}) = h_i(x_o) = 0$ and so $\langle \hat{x} - x_o, \nabla h_i(x_o) \rangle$ for each $i \in E$. We constructed a point $x = \hat{x} - x_o$ such that $\langle x, \nabla g_i(x_o) \rangle < 0$ for every $i \in I$ and $\langle x, \nabla h_i(x_o) \rangle = 0$ for every $i \in E$ and so Cottle qualification holds. iii is obvious. We will pass iv by showing that if Cottle constraint qualification does not hold then neither Karlin qualification nor the independence qualification hold. Suppose that the system $\langle x, \nabla g_i(x_o)\rangle < 0$ for $i \in I$ and $\langle x, \nabla h_i(x_o)\rangle = 0$ for $i \in E$ has no solution. By Lemma 5.1.2 there exist scalars $u_i \geq 0 (i \in I)$ and $v_i (i \in E)$ such that $\sum_{i \in I} u_i \nabla g_i(x_o) + \sum_{i \in E} v_i \nabla h_i(x_o) = 0$ where not all the u_i's and v_i's are zeros. This immediately shows that $\nabla g_i(x_o)$ for $i \in E$ and $\nabla h_i(x_o)$ for $i \in E$ cannot be linearly independent and so the independence qualification does not hold. This shows iii. To show that Karlin qualification does not hold we show that $\sum_{i \in I} u_i g_i(x) + \sum_{i \in E} v_i h_i(x) \geq 0$ for every $x \in X$. So let $x \in X$. By convexity of g_i at x_o and linearity of h_i we have $g_i(x) \geq g_i(x_o) + \langle x - x_o, \nabla g_i(x_o)\rangle$ for every $i \in I$ and $h_i(x) = h_i(x_o) + \langle x - x_o, \nabla h_i(x_o)\rangle$ for every $i \in E$. Multiplying the former inequality by $u_i \geq 0$ and the last equality by v_i and adding over the index sets I and E we get $\sum_{i \in I} u_i g_i(x) + \sum_{i \in E} v_i h_i(x) \geq \sum_{i \in I} u_i g_i(x_o) + \sum_{i \in E} v_i h_i(x_o) +$ $\langle x - x_o, \sum_{i \in I} u_i \nabla g_i(x_o) + \sum_{i \in E} v_i \nabla h_i(x_o)\rangle$. Noting that $g_i(x_o) = 0$ for every $i \in I$, $h_i(x_o) = 0$ for each $i \in E$, and $\sum_{i \in I} u_i \nabla g_i(x_o) + \sum_{i \in E} v_i \nabla h_i(x_o) = 0$ we conclude that $\sum_{i \in I} u_i g_i(x) + \sum_{i \in E} v_i h_i(x) \geq 0$ for each $x \in X$ and so Karlin qualification does not hold.

Finally to prove v, So suppose that the reverse constraint qualification holds. In other words assume that $x_o \in \text{int } X$, $g_i (i \in I)$ are pseudo-concave at x_o, and $h_i (i \in E)$ are pseudo-linear at x_o. We will show that $C \subset F(S, x_o)$. Let $x \in C$, i.e., $\langle x, \nabla g_i(x_o)\rangle \leq 0$ for each $i \in I$ and $\langle x, \nabla h_i(x_o)\rangle = 0$ for each $i \in E$. Consider points of the form $x_o + \lambda x$ where $\lambda > 0$. By pseudo-concavity of g_i at x_o it follows that $\langle x_o + \lambda x - x_o, \nabla g_i(x_o)\rangle = \lambda \langle x, \nabla g_i(x_o)\rangle \leq 0$ implies that $g_i(x_o + \lambda x) \leq g_i(x_o) = 0$ for each $\lambda > 0$ and each $i \in I$. By pseudo-linearity of h_i it follows by an argument similar to the above that $0 \leq h_i(x_o + \lambda x) \leq 0$, i.e., $h_i(x_o + \lambda x) = 0$ for each $\lambda > 0$ and each $i \in E$. So if we choose $\lambda > 0$ sufficiently small it follows by continuity of $g_i (i \notin I)$ at x_o that $g_i(x_o + \lambda x) < 0$. Also since $x_o \in \text{int } X$ then $x_o + \lambda x \in X$ for λ sufficiently small. In other words $x_o + \lambda x \in S$ for all $\lambda > 0$ sufficiently small and hence $x \in F(S, x_o) \subset A(S, x_o)$. This shows that $C \subset A(S, x_o)$ and so the Kuhn-Tucker constraint qualification holds.

Figure 6.4 below summarizes Theorem 6.2.3 above where the relationships among the constraint qualifications are depicted.

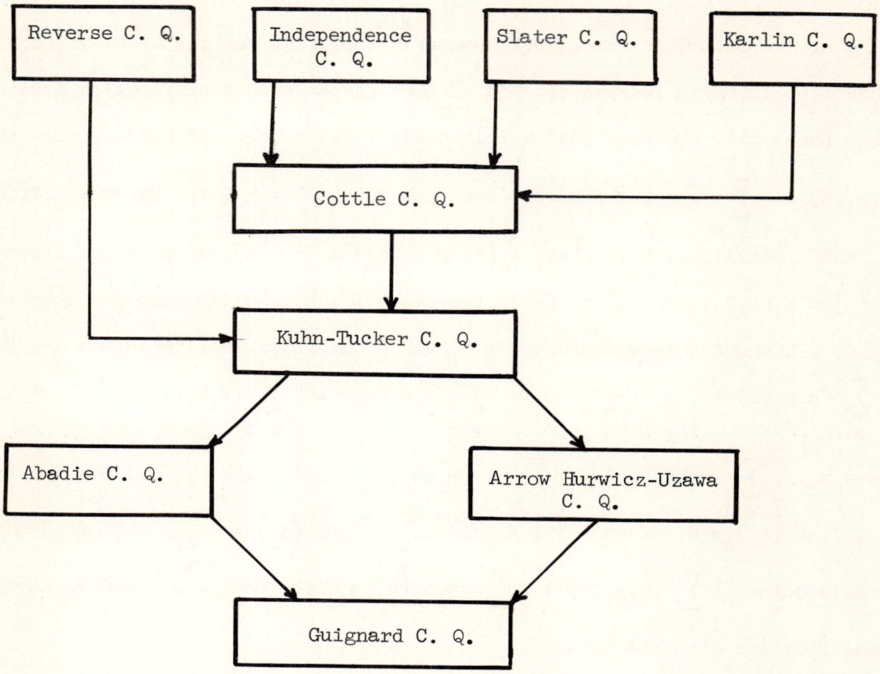

Figure 6.4

Relationships Between Constraint Qualifications For Equality and Inequality Constraint Problems

6.2.4 Theorem. Consider the problem: minimize $f(x)$ subject to $x \in S$ where $S = \{x \in X : g(x) \leq 0 \text{ and } h(x) = 0\}$. Suppose that x_o solves the problem. Further suppose that Guignard constraint qualification $C \subset T^{**}(S,x_o)$ holds. Then there exist scalars $u_i \geq 0$ $(i \in I)$ and $v_i (i \in E)$ such that $\nabla f(x_o) + \sum_{i \in I} u_i \nabla g_i(x_o) + \sum_{i \in E} v_i \nabla h_i(x_o) = 0$.

<u>Proof</u>: By Lemma 6.1.3 $-\nabla f(x_o) \in T^*(S,x_o)$. But since $C \subset T^{**}(S,x_o)$ then $C^* \supset T^*(S,x_o)$ and hence $-\nabla f(x_o) \in C^*$. Note that $C = (\bigcap_{i \in I} C_i) \cap (\bigcap_{i \in E} C_i^-) \cap (\bigcap_{i \in E} C_i^+)$ where $C_i = \{x : \langle x, \nabla g_i(x_o)\rangle \le 0\}$ for $i \in I$, $C_i^- = \{x : \langle x, \nabla h_i(x_o)\rangle \le 0\}$ for $i \in E$ and $C_i^+ = \{x : \langle x, -\nabla h_i(x_o)\rangle \le 0\}$ for $i \in E$. It is obvious that C is a polyhedral cone since it is the intersection of a finite number of halfspaces. So $C^* = \sum_{i \in I} C_i^* + \sum_{i \in E} C_i^{-*} + \sum_{i \in E} C_i^{+*}$. However $C_i^* = \{\lambda \nabla g_i(x_o) : \lambda \ge 0\}$ for $i \in I$, $C_i^{-*} = \{\lambda \nabla h_i(x_o) : \lambda \ge 0\}$ for $i \in E$ and $C_i^{+*} = \{\lambda \nabla h_i(x_o) : \lambda \le 0\}$. It is then obvious that there exist scalars $u_i \ge 0$ for $i \in I$ and v_i for $i \in E$ such that $-\nabla f(x_o) = \sum_{i \in I} u_i \nabla g_i(x_o) + \sum_{i \in E} v_i \nabla h_i(x_o)$ and the proof is complete.

6.3. Necessary and Sufficient Qualification

We have shown in the previous sections that if $T^*(S,x_o) = C^*$ then any objective function f having a minimum (local minimum) over S at x_o satisfies the condition:

$$\nabla f(x_o) + \sum_{i \in I} u_i \nabla g_i(x_o) + \sum_{i=1}^{k} v_i \nabla h_i(x_o) = 0$$

$$u_i \ge 0 \qquad \text{for} \qquad i \in I, \quad \text{where } I = \{i : g_i(x_o) = 0\}.$$

We will now show that $T^*(S,x_o) = C^*$ is the weakest possible qualification that will guarantee that the above condition will hold at x_o (see [18]). In other words, for any given objective function f with a local minimum over S at x_o, if $\nabla f(x_o) + \sum_{i \in I} u_i \nabla g_i(x_o) + \sum_{i=1}^{k} v_i \nabla h_i(x_o) = 0$ where $u_i \ge 0$ for each i, we can claim that $T^*(S,x_o) = C^*$. To prove this we only need to show that $T^*(S,x_o) \subset C^*$ since the reverse inclusion is always true (see Lemma 6.2.1). We will show that for each $y \in T^*(S,x_o)$ there corresponds an objective function f which is differentiable at x_o and has a local minimum over S at x_o with the property that $y = -\nabla f(x_o)$. But from the Kuhn-Tucker necessary conditions we have $-\nabla f(x_o) = \sum_{i \in I} u_i \nabla g_i(x_o) + \sum_{i=1}^{k} v_i \nabla h_i(x_o)$ where $u_i \ge 0$ for every $i \in I$. Hence $y \in C^*$. In order to construct the function we need the following lemma.

6.3.1 Lemma. Let y be a nonzero element in $T^*(S,x_o)$. Let $C_\ell = \{x \in E_n : \langle x - x_o, y\rangle \le \frac{\|x - x_o\|}{\ell\|y\|}\}$, $\ell = 1,2,\ldots$. Then for each ℓ there is an $\varepsilon(\ell) > 0$ such that $S \cap N_{\varepsilon(\ell)} \subset C_\ell$ where $N_{\varepsilon(\ell)} = \{x : \|x - x_o\| < \varepsilon(\ell)\}$.

Proof: Suppose otherwise. Then for some ℓ there exists a sequence $\{x_p \in S \cap N_{i/p}\}$ such that $x_p \notin C_\ell$. This then implies that $\langle x_p - x_o, y\rangle > \frac{\|x_p - x_o\|}{\ell\|y\|}$ for each p, i.e., $\langle \frac{x_p - x_o}{\|x_p - x_o\|}, y\rangle > \frac{1}{\ell\|y\|} > 0$ for each p. But the sequence $\{\frac{x_p - x_o}{\|x_p - x_o\|}\}$ is bounded and hence has a converging subsequence $\frac{x_{p_i} - x_o}{\|x_{p_i} - x_o\|} \to z$. So far we have constructed a sequence $x_{p_i} \to x_o$, $x_{p_i} \in S$, such that $\frac{x_{p_i} - x_o}{\|x_{p_i} - x_o\|} \to z$. By definition of the cone of tangents it is immediate that $z \in T(S,x_o)$. But $\langle \frac{x_{p_i} - x_o}{\|x_{p_i} - x_o\|}, y\rangle > \frac{1}{\ell\|y\|}$ for each i and hence $\langle z,y\rangle \ge \frac{1}{\ell\|y\|} > 0$, which contradicts the fact that $y \in T^*(S,x_o)$. This completes the proof.

We will now give the main theorem which, in effect, shows that Guignard qualification is both necessary and sufficient for the validation of the Kuhn-Tucker conditions.

6.3.2 Theorem. Let $S = \{x \in X : g(x) \le 0,\ h(x) = 0\}$. Suppose that for each objective function f with a local minimum at x_o the following Kuhn-Tucker conditions hold:

$$\nabla f(x_o) + \sum_{i \in I} u_i \nabla g_i(x_o) + \sum_{i=1}^{k} v_i \nabla h_i(x_o) = 0$$

$$u_i \ge 0 \qquad \text{for} \qquad i \in I$$

where $I = \{i : g_i(x_o) = 0\}$. Then $C^* = T^*(S,x_o)$, i.e. Guignard constraint qualification must hold.

Proof: Since $C^* \subset T^*(S,x_o)$ always hold, it suffices to show that $T^*(S,x_o) \subset C^*$. For a given nonzero $y \in T^*(S,x_o)$ we will show that there is an objective function f differentiable at x_o with $-\nabla f(x_o) = y$ and which has a local minimum at x_o. From the Kuhn-Tucker conditions, we then have $-\nabla f(x_o) = \sum_{i \in I} u_i \nabla g_i(x_o) + \sum_{i=1}^{k} v_i \nabla h_i(x_o)$ where $u_i \geq 0$ for each $i \in I$. Therefore $y \in C^*$ implying $T^*(S, x_o) \subset C^*$.

Let $\hat{\epsilon}_\ell = \sup\{\epsilon : S \cap N_\epsilon \subset C_k\}$ where C_ℓ is as defined in Lemma 6.3.1. Now by Lemma 6.3.1 $\hat{\epsilon}_\ell > 0$ for each $\ell \geq 1$. Let

$$\epsilon_1 = \min(1, \hat{\epsilon}_1) \quad \text{and} \quad \epsilon_\ell = \min(\tfrac{1}{2}\,\epsilon_{\ell-1}, \hat{\epsilon}_\ell), \qquad \ell > 1.$$

It is clear that $\epsilon_\ell > \epsilon_{\ell+1}$ for each $\ell \geq 1$, and $\epsilon_\ell \leq (\frac{1}{2})^{\ell-1}$ and hence $\epsilon_\ell \to 0$. We will now define a real valued function α as follows.

$$\alpha(z) = \begin{cases} 2\,\|z\| & \text{for } \|z\| \geq \epsilon_2 \\ 2\,\dfrac{\|z\|}{\ell-1}\,\dfrac{\|z\| - \epsilon_{\ell+1}}{\epsilon_\ell - \epsilon_{\ell+1}} + 2\,\dfrac{\|z\|}{\ell}\,\dfrac{\epsilon_\ell - \|z\|}{\epsilon_\ell - \epsilon_{\ell+1}} & \text{for } \|z\| \in [\epsilon_{\ell+1}, \epsilon_\ell],\ \ell \geq 2 \\ 0 & \text{for } \|z\| = 0 \end{cases}$$

We will show that the function α defined above is differentiable at $z = 0$ and also that $\nabla\alpha(0) = 0$. Note that for $z \in [\epsilon_{\ell+1}, \epsilon_\ell]$, $\ell \geq 2$, the value of α is determined as a convex combination of $2\,\frac{\|z\|}{\ell-1}$ and $\frac{2\|z\|}{\ell}$. Hence it is obvious that α decreases rapidly as $z \to 0$ (since the index ℓ also increases). Now if $\|z\| < \epsilon_\ell$, then

$$\begin{aligned} \alpha(z) &\leq \frac{2\|z\|}{\ell-1}\,\frac{\|z\| - \epsilon_{\ell+1}}{\epsilon_\ell - \epsilon_{\ell+1}} + \frac{2\|z\|}{\ell}\,\frac{\epsilon_\ell - \|z\|}{\epsilon_\ell - \epsilon_{\ell+1}} \\ &< \frac{2\|z\|}{\ell-1}\,\frac{\|z\| - \epsilon_{\ell+1}}{\epsilon_\ell - \epsilon_{\ell+1}} + \frac{2\|z\|}{\ell-1}\,\frac{\epsilon_\ell - \|z\|}{\epsilon_\ell - \epsilon_{\ell+1}} \\ &= \frac{2\|z\|}{\ell-1} \end{aligned}$$

Hence for each $\|z\| < \epsilon_\ell$ we have $0 \leq \alpha(z) < \frac{2\|z\|}{\ell-1}$. Dividing by $\|z\|$, we get $0 \leq \frac{\alpha(z)}{\|z\|} < \frac{2}{\ell-1}$. If we now let $z \to 0$ (i.e., if we let $\ell \to \infty$), we get $0 \leq \lim_{z \to 0} \frac{\alpha(z)}{\|z\|} \leq \lim_{\ell \to \infty} \frac{2}{\ell-1} = 0$. Hence $\lim_{z \to 0} \frac{\alpha(z)}{\|z\|} = 0$ and hence α is differentiable at

$z = 0$ and moreover $\nabla\alpha(0) = 0$.

Now let $y \in T^*(S,x_o)$, $y \neq 0$. For any $x \in E_n$ let $z_x = (x - x_o) - \langle x - x_o, y\rangle y$. We will show that there is a neighborhood N_ϵ about x_o such that for any $x \in S \cap N_\epsilon$, $x \neq x_o$, $\langle x - x_o, y\rangle \|y\| < \alpha(z_x)$. To show this, consider the following cases.

(i) $\langle x - x_o, y\rangle < 0$. In this case the result is immediate since by construction $\alpha(z) \geq 0$ for all z.

(ii) $\langle x - x_o, y\rangle = 0$. In this case $z_x = x - x_o$, and $\|z_x\| = \|x - x_o\| > 0$ since $x \neq x_o$. Then by construction of α, $\alpha(z_x) > 0$. Hence $\alpha(z_x) > \langle x - x_o, y\rangle \|y\|$.

(iii) $\langle x - x_o, y\rangle > 0$. By definition of z_x,

$$\|x - x_o\| - \langle x - x_o, y\rangle\|y\| \leq \|z_x\| \leq \|x - x_o\| + \langle x - x_o, y\rangle\|y\|.$$

Now if $x \in C_\ell$ then $\langle x - x_o, y\rangle \leq \dfrac{\|x - x_o\|}{\ell\|y\|}$ and hence we conclude from the above inequality that $\|x - x_o\| - \frac{1}{\ell}\|x - x_o\| \leq \|z_x\| \leq \|x - x_o\| + \frac{1}{\ell}\|x - x_o\|$, i.e., $\frac{\ell-1}{\ell}\|x - x_o\| \leq \|z_x\| \leq \frac{\ell+1}{\ell}\|x - x_o\|$. We will call this inequality the main inequality. Now suppose that $x \in S \cap N_\epsilon$ for any $0 < \epsilon < \epsilon_3$ then $\|x - x_o\| \in (\epsilon_{\ell+1}, \epsilon_\ell)$ for some $\ell \geq 3$and by Lemma 6.3.1 $S \cap N_\epsilon \subset C_\ell$. But this implies that $\langle x - x_o, y\rangle \leq \dfrac{\|x - x_o\|}{\ell\|y\|} \leq \dfrac{1}{(\ell-1)\;\|y\|}\|z_x\|$ by the main inequality. But also by the main equality we have $\|z_x\| \geq \frac{\ell-1}{\ell}\|x - x_o\| > \frac{\ell-1}{\ell}\epsilon_{\ell+1} > \epsilon_{\ell+2}$. By construction of α, and since $\|z_x\| > \epsilon_{\ell+2}$ it then follows that $\alpha(z_x) > \dfrac{2\|z_x\|}{\ell+1}$. So far we have shown that $\langle x - x_o, y\rangle\|y\| \leq \frac{1}{\ell-1}\|z_x\| < \frac{\ell+1}{2(\ell-1)}\alpha(z_x) \leq \alpha(z_x)$, i.e., for each $x \in S \cap N_\epsilon$ with $0 < \epsilon < \epsilon_3$ it is true that $\langle x - x_o, y\rangle\|y\| < \alpha(z_x)$ as long as $x \neq x_o$.

From the above three cases it is immediate then that there is a neighborhood N_ϵ about x_o such that for any $x \in S \cap N_\epsilon$, $x \neq x_o$, $\langle x - x_o, y\rangle\|y\| < \alpha(z_x)$. We may also note that N_ϵ can be taken as E_n for cases (i) and (ii) and in case (iii) ϵ should be less than ϵ_3. Now we will define the function f as follows: $f(x) = \langle x - x_o, y\rangle - \frac{1}{\|y\|}\alpha(z_x)$. This immediately implies that f has a (unique) minimum over $S \cap N_\epsilon$ at x_o. Also $\nabla f(x_o) = y$ since $z_{x_o} = 0$ and $\nabla\alpha(0) = 0$ as shown earlier. This completes the proof.

CHAPTER 7

CONVEX PROGRAMMING WITHOUT DIFFERENTIABILITY

In this chapter we will develop optimality conditions for nonlinear convex programs. These conditions are of the saddle type where no differentiability assumptions are required. We also develop stationary point optimality conditions in the absence of differentiability, where subgradients of the convex functions play the role of the gradients.

7.1. Saddle Point Optimality Criteria

In this section we will consider a problem of the form minimize $f(x)$ subject to $x \in X$, $g(x) \leq 0$, and $h(x) = 0$. Here X is a convex set in E_n, f is a convex real valued function, $g = (g_1, g_2, \ldots, g_m)$ is a convex function, i.e., each g_i is convex, $h = (h_1, h_2, \ldots, h_k)$ is linear. Clearly if we let $S = \{x \in X : g(x) \leq 0,\ h(x) = 0\}$ then S is a convex set.

The saddle point optimality conditions are of the following form. If x_o solves the above problem then there must exist a nonzero vector $(u_o, \bar{u}, \bar{v})$ such that $u_o \geq 0$, $\bar{u} \geq 0$ and

$$u_o f(x_o) + \langle u, g(x_o)\rangle + \langle v, h(x_o)\rangle \leq u_o f(x_o) + \langle \bar{u}, g(x_o)\rangle + \langle \bar{v}, h(x_o)\rangle$$

$$\leq u_o f(x) + \langle \bar{u}, g(x)\rangle + \langle \bar{v}, h(x)\rangle$$

for all $x \in X$, $u \geq 0$, and $v \in E_k$.

u_o, the lagrangian multiplier associated with f, is not necessarily positive. Under similar conditions to those discussed in Chapter 5, $u_o > 0$. These are called constraint qualifications and will be discussed in more detail later.

In order to develop saddle point optimality conditions for the above problem it will be convenient to use the more general notion of cones and convexity of a function with respect to a cone. So we consider the following definition of cone convexity.

7.1.1 Definition. Let $\beta : X \to E_m$ where X is a convex set in E_n. Let K be a convex cone in E_m. β is said to be K-convex (convex with respect to K) if for each $x_1, x_2 \in X$, $\beta(\lambda x_1 + (1-\lambda)x_2) - \lambda\beta(x_1) - (1-\lambda)\beta(x_2) \in K$ for each $\lambda \in (0,1)$.

From the above definition it is clear that β is convex on X in the ordinary sense if K is the nonpositive orthant in E_m. Also if K is the zero cone in E_m then β is linear on X.

We will now present a theorem of the alternative which will help us develop the saddle point optimality criteria.

7.1.2 <u>Theorem</u>. Let X be a nonempty convex set in E_n and let $\alpha : X \to E_1$ be convex on X. Let $\beta : X \to E_m$ be K-convex where K is a convex cone in E_m. If the system

$$\alpha(x) < 0 \quad \beta(x) \in K \quad \text{has no solution in X}$$

then the system

$$p\,\alpha(x) + \langle q, \beta(x) \rangle \geq 0 \quad \text{for all } x \in X$$

$$p \geq 0 \quad q \in K^* \quad (p,q) \neq (0,0)$$

has a solution $(p,q) \in E_{m+1}$. The converse holds if $p > 0$.

<u>Proof</u>: Suppose that the first system has no solution and consider the following set.

$$\Lambda = \{(y,z) : (y,z) \in E_1 \times E_m, \quad y > \alpha(x) \quad \beta(x) - z \in K \text{ for some } x \in X\}.$$

It can be easily verified that Λ is convex and $(0,0) \notin \Lambda$. Then from Theorem 2.4.12 there exists a non-zero vector (p,q) such that $\langle (p,q), (y,z) \rangle \geq 0$ for each $(y,z) \in \Lambda$, i.e., $py + \langle q,z \rangle \geq 0$ whenever $y > \alpha(x)$ and $\beta(x) - z \in K$ for some $x \in X$. Now consider an arbitrary fixed element x of X and let ξ be an arbitrary fixed element of K. Consider (y_ϵ, z_λ) where

$$y_\epsilon = \alpha(x) + \epsilon \quad \text{and} \quad z_\lambda = \beta(x) - \lambda\xi$$

For $\lambda > 0$, $\beta(x) - z_\lambda = \lambda\xi \in K$. Hence from the definition of Λ, we have $(y_\epsilon, z_\lambda) \in \Lambda$ for each $\epsilon > 0$ and $\lambda > 0$. Hence $0 \leq (py_\epsilon + \langle q, z_\lambda \rangle) = p\alpha(x) + p\,\epsilon + \langle q, \beta(x) \rangle - \langle q, \lambda\xi \rangle$. The remaining parts of the theorem are immediate as follows:

(i) Since ϵ can be made as large as we wish, we must have $p \geq 0$ for the above inequality to hold for a fixed $x \in X$, $\xi \in K$, and a fixed λ.

(ii) Since λ can be chosen as large as we wish, we must have $\langle q, \xi \rangle \leq 0$ for the above inequality to hold for a fixed $x \in X$, $\xi \in K$ and a fixed ϵ. But since this is true for any $\xi \in K$ then $q \in K^*$.

(iii) By letting $\epsilon, \lambda \to 0^+$, we get $0 \le p\alpha(x) + \langle q, \beta(x)\rangle$.

Therefore the second system has a solution if the first system has no solution. Now suppose that the second system has a solution with $p > 0$, i.e., there exists a $p > 0$, $q \in K^*$ such that $p\alpha(x) + \langle q, \beta(x)\rangle \ge 0$ for each $x \in X$. Now if $\beta(x) \in K$, then since $q \in K^*$, we have $\langle q, \beta(x)\rangle \le 0$. Hence $p\alpha(x) \ge 0$ which implies that $\alpha(x) \ge 0$ since $p > 0$. Thus $\beta(x) \in K$ and $\alpha(x) < 0$ has no solution. This completes the proof.

It is worthwhile noting that if K is an open cone then K-convexity of β can be replaced by the weaker assumption of $C\ell$ K-convexity of β.

Theorem 7.1.3 below gives the saddle point optimality criteria. The problem under consideration is to minimize $f(x)$ subject to $x \in X$ and $\beta(x) \in K$.

7.1.3 Theorem. Suppose that x_o solves the problem minimize $f(x)$ subject to $x \in X$ and $\beta(x) \in K$. Here X is a convex set, f is convex, K is a convex cone, and β is K-convex. Then there exists a nonzero $(u_o, \bar{u})$ such that $u_o \ge 0$, $\bar{u} \in K^*$ such that:

$$u_o f(x_o) + \langle u, \beta(x_o)\rangle \le u_o f(x_o) + \langle \bar{u}, \beta(x_o)\rangle \le u_o f(x) + \langle \bar{u}, \beta(x)\rangle$$

for all $x \in X$ and $u \in K^*$.

Proof: If x_o solves the above problem then the system $\alpha(x) < 0$, $\beta(x) \in K$ has no solution in X where $\alpha(x) = f(x) - f(x_o)$. Then by Theorem 7.1.2 above there exists a nonzero $(u_o, \bar{u})$ such that $u_o \ge 0$, $\bar{u} \in K^*$ and $u_o[f(x) - f(x_o)] + \langle \bar{u}, \beta(x)\rangle \ge 0$ for all $x \in X$. By letting $x = x_o$ we get $\langle \bar{u}, \beta(x_o)\rangle \ge 0$. On the other hand $\beta(x_o) \in K$ and $\bar{u} \in K^*$ implies $\langle \bar{u}, \beta(x_o)\rangle \le 0$. Hence $\langle \bar{u}, \beta(x_o)\rangle = 0$. From the above inequality and the fact $\langle \bar{u}, \beta(x_o)\rangle = 0$ we must have $u_o f(x) + \langle \bar{u}, \beta(x)\rangle \ge u_o f(x_o) = u_o f(x_o) + \langle \bar{u}, \beta(x_o)\rangle$ for all $x \in X$. To complete the proof note that $u \in K^*$ implies that $\langle u, \beta(x_o)\rangle \le 0 = \langle \bar{u}, \beta(x_o)\rangle$ and hence $u_o f(x_o) + \langle u, \beta(x_o)\rangle \le u_o f(x_o) + \langle \bar{u}, \beta(x_o)\rangle$ for all $u \in K^*$.

Corollary: Let X be a convex set, let f and g be convex, and let h be linear. If x_o solves the problem minimize $f(x)$ subject to $x \in X$, $g(x) \le 0$, $h(x) = 0$ then there exists a nonzero $(u_o, \bar{u}, \bar{v})$ such that $u_o \ge 0$, $\bar{u} \ge 0$ and

$$u_o f(x_o) + \langle u, g(x_o)\rangle + \langle v, h(x_o)\rangle \le u_o f(x_o) + \langle \bar{u}, g(x_o)\rangle + \langle \bar{v}, h(x_o)\rangle$$

$$\le u_o f(x) + \langle \bar{u}, g(x)\rangle + \langle \bar{v}, h(x)\rangle$$

for all $x \in X$ and all $u \geq o$. In particular the complementary slackness condition $\bar{u}_i g_i(x_o) = 0$ for $i = 1,2,\ldots,m$ holds.

Proof: Let $\beta = (g,h)$ and let K be the cross product of the nonpositive orthant and the zero vector. Then apply the above theorem. Also note that $\langle \bar{u}, \beta(x_o) \rangle = 0$.

By requiring differentiability at one point x_o, the Fritz John conditions follow as shown below.

7.1.4 Theorem. Suppose that x_o solves the problem minimize $f(x)$ subject to $x \in X$, $g(x) \leq 0$, and $h(x) = 0$. Here X is a convex set in E_n, $g = (g_1, g_2, \ldots, g_m)$ is convex, $h = (h_1, h_2, \ldots, h_k)$ is linear. Furthermore f and g are differentiable at x_o. Then there exist a nonzero vector $(u_o, u_1, u_2, \ldots, u_m, v_1, v_2, \ldots, v_k)$ such that:

i. $u_o \geq 0 \quad u_i \geq 0, \quad u_i g_i(x_o) = 0 \quad$ for $i = 1,2,\ldots,m$.

ii. $\langle x - x_o, u_o \nabla f(x_o) + \sum_{i=1}^{m} u_i \nabla g_i(x_o) + \sum_{i=1}^{k} v_i \nabla h_i(x_o) \rangle \geq 0$ for all $x \in X$.

Furthermore if $x_o \in \text{int } X$ then ii reduces to $u_o \nabla f(x_o) + \sum_{i=1}^{m} u_i \nabla g_i(x_o) + \sum_{i=1}^{k} v_i \nabla h_i(x_o) = 0$.

Proof: Since x_o solves the above problem then by the corollary to Theorem 7.1.3 there is a nonzero $(u_o, u_1, \ldots, u_m, v_1, \ldots, v_k)$ such that $u_o \geq 0$, $u_i \geq 0$, $u_i g_i(x_o) = 0$ for $i = 1,2,\ldots,m$ and $u_o f(x_o) + \sum_{i=1}^{m} u_i g_i(x_o) + \sum_{i=1}^{k} v_i h_i(x_o) \geq u_o f(x) + \sum_{i=1}^{m} u_i g_i(x) + \sum_{i=1}^{k} v_i h_i(x)$ for each $x \in X$. Let $x \in X$. Then $x_o + \lambda(x - x_o) \in X$ for each $\lambda \in (0,1)$ by convexity of X. This means that $u_o[f(x_o) - f(x_o + \lambda(x - x_o))] + \sum_{i=1}^{m} u_i[g_i(x_o) - g_i(x_o + \lambda(x - x_o))] + \sum_{i=1}^{m} v_i[h_i(x_o + \lambda(x - x_o))] \geq 0$ for each $\lambda \in (0,1)$. Dividing by $\lambda > 0$, and taking the limit as $\lambda \to 0^+$ we get $\langle x - x_o, u_o \nabla f(x_o) + \sum_{i=1}^{m} u_i \nabla g_i(x_o) + \sum_{i=1}^{k} v_i \nabla h_i(x_o) \rangle \geq 0$. To complete the proof let us investigate the case when $x_o \in \text{int } X$. Since $x_o \in \text{int } X$ then we can pick a $\lambda > 0$ such that $x_o - \lambda[u_o \nabla f(x_o) + \sum_{i=1}^{m} u_i \nabla g_i(x_o) + \sum_{i=1}^{k} v_i \nabla h_i(x_o)] \in X$. Substituting this in the above inequality it follows that $-\lambda \| u_o \nabla f(x_o) + \sum_{i=1}^{m} u_i \nabla g_i(x_o) + \sum_{i=1}^{k} v_i \nabla h_i(x_o) \|^2 \geq 0$ which implies that $u_o \nabla f(x_o) + \sum_{i=1}^{m} u_i \nabla g_i(x_o) + \sum_{i=1}^{k} v_i \nabla h_i(x_o) = 0$.

Sufficiency of the Saddle Point Conditions

It may be noted that if $u_o > 0$ and the saddle point criteria hold then x_o indeed solves the minimization problem. In other words the saddle point criteria are sufficient for the point under investigation to solve the nonlinear program, provided of course that $u_o > 0$. Here no convexity assumptions of any kind are required. This result is given by Theorem 7.1.5 below. Note that this theorem follows directly from the converse part of Theorem 7.1.2.

7.1.5 Theorem. Let $x_o \in X$ and suppose that there exists a vector $(\bar{u}_1, \bar{u}_2, \ldots, \bar{u}_m, \bar{v}_1, \bar{v}_2, \ldots, \bar{v}_k)$ where $\bar{u}_i \geq 0$ for $i = 1,2,\ldots,m$ such that

$$f(x_o) + \sum_{i=1}^{m} u_i g_i(x_o) + \sum_{i=1}^{k} v_i h_i(x_o) \leq f(x_o) + \sum_{i=1}^{m} \bar{u}_i g_i(x_o) + \sum_{i=1}^{k} \bar{v}_i h_i(x_o)$$

$$\leq f(x) + \sum_{i=1}^{m} \bar{u}_i g_i(x) + \sum_{i=1}^{k} \bar{v}_i h_i(x)$$

for all $x \in X$, all nonnegative $u = (u_1, u_2, \ldots, u_m)$, and all arbitrary $v = (v_1, v_2, \ldots, v_k)$. Then x_o solves the problem minimize $f(x)$ subject to $x \in X$, $g(x) \leq 0$ and $h(x) = 0$.

Proof: We will first show that x_o is a feasible solution. From the inequality $\sum_{i=1}^{m} u_i g_i(x_o) + \sum_{i=1}^{k} v_i h_i(x_o) \leq \sum_{i=1}^{m} \bar{u}_i g_i(x_o) + \sum_{i=1}^{k} \bar{v}_i h_i(x_o)$ for all $u \geq 0$ and all $v \in E_k$, it is clear that $g_i(x_o) \leq 0$ for $i = 1,2,\ldots,m$ and $h_i(x_o) = 0$ for $i = 1,2,\ldots,k$. Since $x_o \in X$, $g(x_o) \leq 0$, and $h(x_o) = 0$ then x_o is a feasible solution to the nonlinear program. By letting u = 0 in the first part of the saddle inequality and since $h(x_o) = 0$ it follows that $\sum_{i=1}^{m} \bar{u}_i g_i(x_o) \geq 0$. Since $\bar{u}_i \geq 0$ and $g_i(x_o) \leq 0$ for each i, then $\sum_{i=1}^{m} \bar{u}_i g_i(x_o) = 0$. Therefore the second part of the saddle inequality implies that $f(x_o) \leq f(x) + \sum_{i=1}^{m} \bar{u}_i g_i(x) + \sum_{i=1}^{k} \bar{v}_i h_i(x)$ for each $x \in X$. Now let x be a feasible solution of the nonlinear program, i.e., $x \in X$, $g_i(x) \leq 0$ for each i and $h_i(x) = 0$ for each i. Since $\bar{u}_i \geq 0$ and $g_i(x) \leq 0$ then $\bar{u}_i g_i(x) \leq 0$ and we get $f(x_o) \leq f(x)$. This shows that x_o is indeed an optimal solution of the problem.

Constraint Qualifications

We will now briefly discuss some conditions (constraint qualifications) which guarantee that $u_o > 0$. Under these qualifications one gets Kuhn-Tucker type saddle point optimality conditions.

Let us consider nonlinear programs of the form minimize f(x) subject to $x \in X$, $g(x) \leq 0$. We will discuss Karlin constraint qualification, Slater constraint qualification, as well as the strict constraint qualification. These are similar to the corresponding qualifications discussed in Chapter 6.

Karlin Constraint Qualification

Let X be a convex set in E_n and let $g = (g_1, g_2, \ldots, g_m)$ be convex. The feasible set $S = \{x \in X : g(x) \leq 0\}$ is said to satisfy Karlin qualification at $x_o \in S$ if there exists no nonzero nonnegative u such that $\langle u, g(x) \rangle \geq 0$ for all $x \in X$.

Slater Constraint Qualification

Let X be a convex set in E_n and let $g = (g\ , g_2, \ldots, g_m)$ be convex. The feasible region $S = \{x \in X : g(x) \leq 0\}$ is said to satisfy Slater constraint qualification at $x_o \in S$ if there exists an $x \in X$ such that $g(x) < 0$.

Strict Constraint Qualification

Let X be convex in E_n and let $g = (g_1, g_2, \ldots, g_m)$ be convex. Then $S = \{x \in X : g(x) \leq 0\}$ is said to satisfy the strict constraint qualification at $x_o \in S$ if S contains two distinct points x_1 and x_2, and g is strictly convex at, say, x_1.

The following lemma shows that Karlin and Slater constraint qualifications are equivalent. It also shows that the strict constraint qualification implies both Karlin and Slater qualifications.

7.1.6 Lemma.

i. Karlin constraint qualification and Slater constraint qualification are equivalent.

ii. Strict constraint qualification implies both Karlin and Slater.

Proof: We first show that Slater qualification implies Karlin qualification. Clearly if there exists an $\hat{x} \in X$ with $g(\hat{x}) < 0$ then for any vector $u \geq 0$ with $u \neq 0$ we must have $\langle u, g(\hat{x}) \rangle < 0$. In other words one can never find a nonzero nonnegative u with $\langle u, g(x) \rangle \geq 0$ for each $x \in X$, i.e., Karlin qualification holds. To show that Karlin qualification implies Slater it suffices to prove that if Slater qualification does not hold then Karlin qualification does not hold. So suppose that Slater qualification does not hold, i.e., the system $g(x) < 0$ has no solution in X. We can apply Theorem 7.1.2 by letting $\beta = g$ and K be the negative orthant in E_m. All the hypotheses of the theorem hold (note here we require convexity of g, i.e., convexity of g with respect to $C\ell\, K$, see the discussion below Theorem 7.1.2). Therefore, there exists a nonzero $u \geq 0$ such that $\langle u, g(x) \rangle \geq 0$ for all $x \in X$. This means that Karlin qualification does not hold. This completes the proof of part i. To prove ii suppose that the strict qualification holds, i.e., $x_1 \neq x_2$ are both in S and g is strictly convex at x_1. $x_1, x_2 \in X$ and by convexity of X we must have $\lambda x_1 + (1-\lambda)x_2 \in X$ for each $\lambda \in (0,1)$. Furthermore by strict convexity of g at x_1 we have $g(\lambda x_1 + (1-\lambda)x_2) < \lambda g(x_1) + (1-\lambda)g(x_2) \leq 0$ by feasibility of both x_1 and x_2. This provides at least one $x \in X$ where $g(x) < 0$, i.e., Slater (and hence Karlin) qualification holds. This completes the proof.

Needless to say that under any of the above constraint qualifications $u_o > 0$. This can be shown as follows. Suppose that Karlin qualification holds and suppose that x_o solves the problem: minimize $f(x)$ subject to $x \in X$ and $g(x) \leq 0$ where X is convex, f and g are convex. Then by the corollary to Theorem 7.1.3 we must have a nonzero vector $(u_o, \bar{u}_1, \ldots, \bar{u}_m)$ such that $u_o f(x_o) + \sum_{i=1}^{m} u_i g_i(x_o) \leq u_o f(x_o) \quad +$

$$\sum_{i=1}^{m} \bar{u}_i g_i(x_o) \le u_o f(x) + \sum_{i=1}^{m} \bar{u}_i g_i(x) \text{ for each } x \in X \text{ and each } u = (u_1, u_2, \ldots, u_m)$$

≥ 0. But $\bar{u}_i g_i(x_o) = 0$ for each i, and so if $u_o = 0$ then we get $\sum_{i=1}^{m} \bar{u}_i g_i(x) \ge 0$ for each $x \in X$. This however violates Karlin qualification and so it is impossible for u_o to vanish. In other words $u_o > 0$.

Without loss of generality u_o can be taken as one. In this case one obtains Kuhn-Tucker saddle point conditions of the following form: Let x_o solve the problem minimize $f(x)$ subject to $x \in X$ and $g(x) \le 0$, where X, f, and g are convex. If any of the constraint qualifications discussed above hold, then there must exist a vector u such that:

i. $u_i \ge 0$ for $i = 1,2,\ldots,m$

ii. $u_i g_i(x_o) = 0 \quad i = 1,2,\ldots,m$

iii. $f(x_o) \le f(x) + \sum_{i=1}^{m} u_i g_i(x)$ for each $x \in X$.

So far we have discussed some constraint qualifications involving only inequality constraints. We will now present a constraint qualification which takes care of both equality and inequality constraints.

Uzawa Constraint Qualification

Let X be convex, $g = (g_1, g_2, \ldots, g_m)$ be convex, and $h = (h_1, h_2, \ldots, h_k)$ be linear. Let $S = \{x \in X : g(x) \le 0,\ h(x) = 0\}$ and let $x_o \in S$. Then Uzawa constraint qualification is satisfied at x_o if:

i. $0 \in \text{int } h(X)$ where $h(X) = \{h(x) : x \in X\}$

ii. There exists an $x \in X$ such that $g(x) < 0$ and $h(x) = 0$.

We may think of condition i above as follows. $0 \in \text{int } h(X)$ essentially means that the image of X under h contains neighborhood of the origin. In particular if X contains the origin in E_n as an interior point then indeed $h(x)$ will contain the origin in E_k as an interior point provided that h has full rank. More specifically if $h(x) = Ax - b$ where A is an kxn matrix, then $0 \in \text{int } h(X)$, provided that A is full rank and X contains the origin as an interior point. The reader may note that A has full rank is no restriction because otherwise we may throw away the redundant equations.

The following remark shows that $u_o > 0$ under Uzawa constraint qualification, and hence one obtains the Kuhn-Tucker saddle point criteria at optimal points.

7.1.7 Lemma. Let $S = \{x \in X : g(x) \leq 0, h(x) = 0\}$ where X is convex, g is convex and h is linear. Suppose that $x_o \in S$ solves the problem: minimize $f(x)$ subject to $x \in X$, $g(x) \leq 0$, and $h(x) = 0$. Further suppose that Uzawa constraint qualification holds. Then there exists a vector $(\bar{u},\bar{v})$ such that:

i. $\bar{u}_i g_i(x_o) = 0, \quad \bar{u}_i \geq 0 \quad i = 1,2,\ldots,m.$

ii. $f(x_o) \leq f(x) + \sum_{i=1}^{m} \bar{u}_i g_i(x) + \sum_{i=1}^{k} \bar{v}_i h_i(x)$ for all $x \in X$

Proof: Since x_o solves the problem then by the corollary to Theorem 7.1.3 there exists a nonzero vector $(u_o,\bar{u},\bar{v})$ such that:

$$\bar{u}_i g_i(x_o) = 0, \quad \bar{u}_i \geq 0 \quad i = 1,2,\ldots,m.$$

$$u_o f(x_o) \leq u_o f(x) + \sum_{i=1}^{m} \bar{u}_i g_i(x) + \sum_{i=1}^{k} \bar{v}_i h_i(x) \quad \text{for all } x \in X.$$

By contradiction suppose that $u_o = 0$. By Uzawa qualification there must exist an $\hat{x} \in X$ with $g(\hat{x}) < 0$ and $h(\hat{x}) = 0$. Therefore we get $\sum_{i=1}^{m} \bar{u}_i g_i(\hat{x}) \geq 0$. Since $g_i(\hat{x}) < 0$ and $\bar{u}_i \geq 0$ for all i, then $\sum_{i=1}^{m} \bar{u}_i g_i(\hat{x}) \geq 0$ is only possible if $\bar{u}_i = 0$ for each i. The saddle point inequality then implies $0 \leq \langle \bar{v},h(x)\rangle$ for all $x \in X$. But since $0 \in \text{int } h(X)$ we can pick an $x \in X$ such that $h(x) = -\lambda\bar{v}$ where $\lambda > 0$. This shows that $0 \leq \langle \bar{v}, -\lambda\bar{v}\rangle = -\lambda\|\bar{v}\|^2$ which implies that $\bar{v} = 0$. In other words if $u_o = 0$ then both $\bar{u}$ and $\bar{v}$ are also zeros and so the vector $(u_o,\bar{u},\bar{v}) = (0,0,0)$. This is impossible, however, and hence $u_o > 0$. Without loss of generality u_o may be assumed equal to one and the proof is complete.

7.2. Stationary Point Optimality Conditions

In this section we will develop stationary point optimality conditions for convex programs. Here subgradients of convex functions (which are not necessarily unique) play the role of the gradient vectors. We start by reviewing the definition of a subgradient of a convex function.

7.2.1 Definition. Let $f : E_n \to E_1$ be a convex function. ξ is said to be a subgradient of f at x_o if $f(x) \geq f(x_o) + \langle x - x_o, \xi \rangle$ for each $x \in E_n$.

It is clear that ξ is a subgradient of f at x_o if and only if $(\xi,-1) \in E_{n+1}$ is a normal to a supporting hyperplane of the epigraph of f at the point $(x_o, f(x_o))$. The reader may note that a convex function indeed has a subgradient at each point in the relative interior of its domain as shown by Theorem 4.1.7.

Now let us consider the problem: minimize $f(x)$ subject to $x \in X$ and $g(x) \leq 0$. Here f is a convex real valued function, $g = (g_1, g_2, \ldots, g_m)$ is a convex vector valued function, and X is a convex set in E_n. Here we suppose that f and g are defined on an open set containing X. Now suppose that $x_o \in \text{int } X$ solves the problem, i.e., $x \in X$ with $g(x) \leq 0$ implies that $f(x) \geq f(x_o)$. Let I be the set of binding constraints, i.e., $I = \{i : g_i(x_o) = 0\}$.

Let us consider the following sets. $\hat{E}_f = \{(x - x_o, y) : y \geq f(x) - f(x_o)\}$, $\hat{E}_{g_i} = \{(x - x_o, y) : y \geq g_i(x)\}$ for each $i \in I$. Note that $\hat{E}_f$ is the epigraph of $f - f(x_o)$ where as $\hat{E}_{g_i}$ is the epigraph of g_i. Note that $\hat{E}_f$ and $\hat{E}_{g_i}$ for each $i \in I$ are convex sets. Denote the intersection of these sets by $\hat{E}$, i.e., $\hat{E} = \hat{E}_f \cap \bigcap_{i \in I} \hat{E}_{g_i}$. The following lemma characterizes the polars of $\hat{E}_f$ and $\hat{E}_{g_i}$. We will use this result in proving Theorem 7.2.3.

7.2.2 Lemma. Let (u,v) be a nonzero element of $(\hat{E}_f)^*$. Then $(u,v) = \lambda(\xi,-1)$ where $\lambda > 0$ and ξ is a subgradient of f at x_o. Similarly for a nonzero $(u,v) \in (\hat{E}_{g_i})^*$ we have $(u,v) = \lambda(\xi,-1)$ where $\lambda > 0$ and ξ is a subgradient of g_i at x_o for each $i \in I$.

Proof: Let (u,v) be a nonzero element of $(\hat{E}_f)^*$. Then $\langle (x - x_o, y), (u,v) \rangle \leq 0$ for each $(x - x_o, y) \in \hat{E}_f$. Since y can be made arbitrarily large then $v \leq 0$. If $v = 0$, then $\langle x - x_o, u \rangle \leq 0$ for all $x \in E_n$. This is only possible if $u = 0$, which in turn implies that $(u,v) = (0,0)$. This violates our assumption. Therefore $v < 0$. Then $(u,v) = \lambda(\xi,-1)$ where $\lambda = -v > 0$ and $\xi = \frac{u}{-v}$. Since $(u,v) \in (\hat{E}_f)^*$ then $\lambda \langle (x - x_o, y), (\xi,-1) \rangle \geq 0$ for all $y \geq f(x) - f(x_o)$. Since $\lambda > 0$ then by letting $y = f(x) - f(x_o)$ we get $f(x) \geq f(x_o) + \langle x - x_o, \xi \rangle$, i.e., ξ is a subgradient of f

at x_o. The case of g_i for $i \in I$ is proved in a similar fashion.

Generalized Fritz John Conditions

At this stage we will develop stationary optimality conditions for convex programs in the absence of differentiability. These conditions can be viewed as a generalization of one Fritz John stationary optimality conditions.

7.2.3 Theorem. Let $x_o \in \text{int } X$ solve the problem: minimize $f(x)$ subject to $x \in X$ and $g(x) \leq 0$. Here X is a convex set in E_n, f and g are real and vector valued convex functions on E_n. Let $I = \{i : g_i(x_o) = 0\}$. Then there exist scalars u_o and u_i $(i \in I)$ not all zero as well a set of vectors ξ and ξ_i $(i \in I)$ such that:

i. $u_o \geq 0 \quad u_i \geq 0 \quad$ for $i \in I$

ii. $u_o\xi + \sum_{i \in I} u_i\xi_i = 0$

iii. ξ is a subgradient of f at x_o

ξ_i is a subgradient of g_i at x_o

Proof: Since $x_o \in \text{int } X$ and since g_i for $i \notin I$ are continuous (see Theorem 4.2.1) then there is a neighborhood N of x_o such that $x \in N$ implies that $x \in X$ and $g_i(x) < 0$ for $i \notin I$. Let $C = \{(x - x_o, y) : x \in N\}$. We will show that $(x - x_o, y) \in \hat{E} \cap C$ implies that $y \geq 0$. Indeed if $(x - x_o, y) \in \hat{E} \cap C$ and x is a feasible solution of the problem then $f(x) \geq f(x_o)$ and so $y \geq f(x) - f(x_o) \geq 0$. On the other hand if $(x - x_o, y) \in \hat{E} \cap C$ where x is not feasible then $g_i(x) > 0$ for some $i \in I$ and so $y \geq g_i(x) > 0$ (note that the possibility $x \notin X$ or $g_i(x) > 0$ for $i \notin I$ are excluded since $x \in N$). At any rate we showed that $(x - x_o, y) \in \hat{E} \cap C$ implies that $y \geq 0$. This shows that $(0,-1) \in (\hat{E} \cap C)^* = (\hat{E}_f \cap (\bigcap_{i \in I} \hat{E}_{g_i}) \cap C)^*$. Note that $\hat{E}_f$, $\hat{E}_{g_i}$ $(i \in I)$ and C are convex sets having interior points in common, e.g., $(0,1) \in \text{int } \hat{E}_f \cap (\bigcap_{i \in I} \text{int } \hat{E}_{g_i}) \cap \text{int } C$. Therefore by Theorem 3.1.4, $(\hat{E}_f \cap (\bigcap_{i \in I} \hat{E}_{g_i}) \cap C)^* = (\hat{E}_f) + \sum_{i \in I} (\hat{E}_{g_i})^* + C^*$. However since $0 \in \text{int } C$ then $C^* = \{0\}$. In other words if x_o solves the nonlinear problem then we must have $(0,-1) \in (\hat{E}_f)^* + \sum_{i \in I} (\hat{E}_{g_i})^*$. However, by Lemma 7.2.2. it immediately follows that $(0,-1) = u_o(\xi,-1) + \sum_{i \in I} u_i(\xi_i,-1)$ where ξ is a subgradient of f at x_o, ξ_i is a subgradient of g_i at x_o, $u_o \geq 0$ and $u_i \geq 0$ for each $i \in I$. Clearly not all the u_i's are zero and the proof is complete.

The conditions given by the above theorem essentially say that if x_o solves the nonlinear program, then one can find subgradients of the objective function and of the binding constraints which are not in a half-space. Consider the problem min $f(x)$ where $f(x) = \max (2x_2 - 3x_1 - 3, - x_2)$ subject to $g_1(x) = x_1^2 + x_2^2 - 1 \leq 0$ and $g_2(x) = x_1 + x_2 - 1 \leq 0$. The optimal point is $x = (0,1)$. We have $\nabla g_1(1,0) = (0,2)$ and $\nabla g_2(1,0) = (1,1)$ and are the only subgradients of g_1 and g_2 respectively at $(1,0)$. Any subgradient of f at $(1,0)$ can be represented as $\lambda(-3,2) + (1-\lambda)(0,-1) = (-3\lambda, 3\lambda-1)$ for $\lambda \in [0,1]$. In particular, for $\lambda = 1/9$, the subgradient is $\xi = (- 1/3, - 2/3)$ which is not contained in any half space containing ∇g_1 and ∇g_2. Here the values of the scalars mentioned in the theorem can easily be verified to be $u_o = 1$, $u_1 = 1/6$ and $u_2 = 1/3$.

So far we showed that the above generalized Fritz John conditions are necessary for optimality. It turns out that these conditions are also sufficient provided that u_o, the lagrangian corresponding to the objective function, is positive. This fact is proved below.

7.2.4 <u>Theorem</u>. Consider the problem: minimize $f(x)$ subject to $x \in X$ and $g(x) \leq 0$. Here X is a convex set in E_n, f and $g = (g_1, g_2, \ldots, g_m)$ are convex functions on E_n. Let x_o be a feasible point and let $I = \{i : g_i(x_o) = 0\}$. Suppose that there exist nonnegative scalars $u_i (i \in I)$ and vectors ξ and ξ_i such that $\xi + \sum_{i \in I} u_i \xi_i = 0$ where ξ and $\xi_i (i \in I)$ are subgradients of f and $g_i (i \in I)$ at x_o respectively. Then x_o solves the above problem.

<u>Proof</u>: Let x be a feasible solution of the problem. We want to show that $f(x) \geq f(x_o)$. Since ξ is a subgradient of f at x_o and ξ_i is a subgradient of g_i at x_o we get:

$$f(x) \geq f(x_o) + \langle x - x_o, \xi \rangle \qquad \text{for each } x \in X$$

$$g_i(x) \geq g_i(x_o) + \langle x - x_o, \xi_i \rangle = \langle x - x_o, \xi_i \rangle \qquad \text{for each } x \in X, \text{ each } i \in I.$$

Multiplying the latter inequality by $u_i \geq 0$, summing over i, and adding the resultant to the first inequality we get

$$f(x) + \sum_{i \in I} u_i g_i(x) \geq f(x_o) + \langle x - x_o, \xi + \sum_{i \in I} u_i \xi_i \rangle \text{ for each } x \in X.$$

Noting that $\xi + \sum_{i \in I} u_i \xi_i = 0$ by hypothesis and that $g_i(x) \leq 0$ for feasible x then the above inequality implies that $f(x) \geq f(x_o)$ and the proof is complete.

Constraint Qualification

In order to insure positivity of u_o we give below two constraint qualifications. Under these qualifications we obtain a generalized form of the Kuhn-Tucker conditions.

1. Slater Constraint Qualification

Let $S = \{x \in X : g(x) \leq 0\}$ where X is convex and $g = (g_1, g_2, \ldots, g_m)$ is convex. Slater Constraint Qualification is satisfied at $x_o \in S$ if there exists an $x \in X$ with $g_i(x) < 0$ for each $i \in I$.

2. Cottle Constraint Qualification

Let $S = \{x \in X : g(x) \leq 0\}$ where X is convex and $g = (g_1, g_2, \ldots, g_m)$ is convex. Cottle Constraint Qualification is satisfied at $x_o \in S$ if given any set of subgradients $\{\xi_i\}_{i \in I}$, the system $\langle y, \xi_i \rangle < 0$ fir $i \in I$ has a solution.

The reader may note that an equivalent form for Cottle qualification is that for any set of subgradients $\{\xi_i\}_{i \in I}$ the system $\sum_{i \in I} u_i \xi_i = 0$ has no nonnegative nonzero solution. Yet another way to put this qualification is that the set of all subgradients of all the binding constraints are contained in an open halfspace.

The following lemma shows that Slater constraint qualification implies Cottle qualification while the latter implies positivity of u_o.

7.2.5 Lemma.

i. Slater constraint qualification implies Cottle constraint qualification.

ii. Both qualifications imply that $u_o > 0$.

Proof: To prove i, suppose that Slater qualification holds at $x_o \in S$. In other words there is a point $x \in X$ with $g_i(x) < 0$ for each $i \in I$. By convexity of g_i it follows that $0 > g_i(x) \geq g_i(x_o) + \langle x - x_o, \xi_i \rangle = \langle x - x_o, \xi_i \rangle$ for each $i \in I$ and each subgradient ξ_i of g_i at x_o. This implies that there exists a vector $y = x - x_o$ such that $\langle y, \xi_i \rangle < 0$ for each $i \in I$ and each set of subgradients. In view of i, to prove ii, it suffices to show that Cottle qualification implies that $u_o > 0$. If x_o solves the nonlinear program: minimize $f(x)$ subject to $x \in X$ and $g(x) \leq 0$ then by

Theorem 7.2.3 there exist nonnegative scalars u_0 and u_i $(i \in I)$ as well as vectors ξ and ξ_i $(i \in I)$ such that $u_0\xi + \sum_{i \in I} u_i\xi_i = 0$ where ξ and ξ_i are subgradients of f and g_i respectively. If $u_0 = 0$ then we have a set of subgradients $\{\xi_i\}_{i \in I}$ as well as nonnegative scalars u_i $(i \in I)$ not all zero with $\sum_{i \in I} u_i\xi_i = 0$, violating Cottle constraint qualification. This shows that $u_0 > 0$ and the proof is complete.

CHAPTER 8

LAGRANGIAN DUALITY

In this chapter we will consider a nonlinear programming problem of the form: minimize $\{f(x) : x \in X, g(x) \in K\}$ where K is an arbitrary set in E_n and K is a closed convex cone. Many of the results of this chapter are applicable to nonconvex and also nondifferentiable functions. The lagrangian duality formulation is obtained by incorporating the constraints in the objective function via the lagrangian multipliers (or dual variables), and hence the name.

8.1. Definitions and Preliminary Results

Consider the following definition of the primal and lagrangian dual problems.

8.1.1 Definition. Let K be a closed convex cone in E_m and X be arbitrary set in E_n. Let $f : E_n \to E_1$ and $g : X \to E_m$ be arbitrary. Then the primal and dual Lagrangian problems are defined by:

Primal: Find $\inf f(x)$

subject to $x \in X$ and $g(x) \in K$

Dual: Find $\sup \theta(u)$

subject to $u \in K^*$

where

$$\theta(u) = \inf_{x \in X} \{f(x) + \langle u, g(x) \rangle\}$$

The following lemma shows that θ is a concave function and hence the dual problem is a concave program. Note that this result holds in the absence of any convexity assumptions of the primal problem.

8.1.2 Lemma. The dual function θ defined by

$$\theta(u) = \inf_{x \in X} \{f(x) + \langle u, g(x) \rangle\}$$

is concave.

Proof: By definition of θ,

$$\begin{aligned}\theta(\lambda u_1 + (1-\lambda)u_2) &= \inf_{x \in X} \{f(x) + \langle(\lambda u_1 + (1-\lambda)u_2), g(x)\rangle\} \\ &= \inf_{x \in X} \{\lambda(f(x) + \langle u_1, g(x)\rangle) + (1-\lambda)(f(x) + \langle u_2, g(x)\rangle)\} \\ &\geq \lambda \inf_{x \in X} \{f(x) + \langle u_1, g(x)\rangle\} + (1-\lambda) \inf_{x \in X} \{f(x) + \langle u_2, g(x)\rangle\} \\ &= \lambda\, \theta(u_1) + (1-\lambda)\theta(u_2) \quad \text{for each } \lambda \in (0,1)\end{aligned}$$

Hence θ is concave.

The following is a weak duality theorem which shows that $f(x) \geq \theta(u)$ for each feasible solution of the primal and dual problems. This means that the dual objective of any $u \in K^*$ is a lower bound on the objective value of all primal feasible solutions.

8.1.3 Theorem.

$$\inf \{f(x) : x \in X,\ g(x) \in K\} \geq \sup \{\theta(u) : u \in K^*\}$$

Proof: Let $x \in X$ and $g(x) \in K$, and let $u \in K^*$. Then we need to show that $f(x) \geq \theta(u)$. But

$$\theta(u) = \inf_{y \in X} \{f(y) + \langle u, g(y)\rangle\} \leq f(x) + \langle u, g(x)\rangle \leq f(x)$$

since $g(x) \in K$ and $u \in K^*$ implies $\langle u, g(x)\rangle \leq 0$.

Some important consequences of the above theorem are:

1. If $\inf \{f(x) : x \in X,\ g(x) \in K\} = -\infty$, then $\theta(u) \equiv -\infty$ for all $u \in K^*$.
2. If $\sup \{\theta(u) : u \in K^*\} = \infty$, then the primal problem is infeasible, i.e., there is no $x \in X$ with $g(x) \in K$.
3. If $f(\bar{x}) = \theta(\bar{u})$ for some $\bar{x} \in \{x : x \in X,\ g(x) \in K\}$ and for some $\bar{u} \in K^*$, then $\bar{x}$ and $\bar{u}$ solve the primal and dual problems respectively.

Perturbation Function. We can perturb the constraints $g(x) \in K$ by a vector z so that the new constraints will be $g(x) - z \in K$. We will now introduce a function of z, called the perturbation function, which will be used to derive certain duality theorems.

8.1.4 Definition. Consider the primal problem of definition 8.1.1. The perturbation function $P : E_m \to E_1$ associated with the problem is defined by

$$P(z) = \inf \{f(x) : x \in X, \quad g(x) - z \in K\}$$

We note that P(0) is precisely the primal problem under consideration. The following lemma show that under suitable convexity assumptions on the primal problem, P is a convex function.

8.1.5 Lemma. Let X be a convex set in E_n and K be a closed convex cone in E_m. Let $f : E_n \to E_1$ be convex and let $g : E_n \to E_m$ be K-convex (see Definition 7.1.1). Then P is a convex function.

Proof: Let $z_1, z_2 \in E_m$. Then for $\lambda \in (0,1)$

$$P(\lambda z_1 + (1-\lambda)z_2) = \inf \{f(x) : x \in X, g(x) - \lambda z_1 - (1-\lambda)z_2 \in K\}.$$

Since $x \in X$ and the latter is a convex set, then we can represent x by $\lambda x_1 + (1-\lambda)x_2$ where $\lambda \in (0,1)$ and $x_1, x_2 \in X$. Then

$$P(\lambda z_1 + (1-\lambda)z_2) = \inf_{x_1,x_2} \{f(\lambda x_1 + (1-\lambda)x_2) : x_1,x_2 \in X, 0 \leq \lambda \leq 1,$$
$$g(\lambda x_1 + (1-\lambda)x_2) - \lambda z_1 - (1-\lambda)z_2 \in K\}.$$

Since g is K-convex we get:

$$g(\lambda x_1 + (1-\lambda)x_2) - \lambda g(x_1) - (1-\lambda)g(x_2) \in K.$$

If $g(x_1) - z_1 \in K$ and $g(x_2) - z_2 \in K$, then by convexity of K we get:

$$\lambda g(x_1) - \lambda z_1 + (1-\lambda)g(x_2) - (1-\lambda)z_2 \in K$$

From the above two set inclusions and since K is a convex cone it follows that:

$$g(\lambda x_1 + (1-\lambda)x_2 - \lambda z_1 - (1-\lambda)z_2 \in K.$$

We have shown that $g(x_1) - z_1 \in K$ and $g(x_2) - z_2 \in K$ imply that $g(\lambda x_1 + (1-\lambda)x_2) - \lambda z_1 - (1-\lambda)z_2 \in K$, and hence:

$$P(\lambda z_1 + (1-\lambda)z_2) \leq \inf_{x_1,x_2} \{f(\lambda x_1 + (1-\lambda)x_2 : x_1 \in X,$$
$$g(x_1) - z_1 \in K: \quad x_2 \in X,$$
$$g(x_2) - z_2 \in K, 0 \leq \lambda \leq 1\}.$$

By convexity of f, $f(\lambda x_1 + (1-\lambda)x_2) \leq \lambda f(x_1) + (1-\lambda)f(x_2)$. We then get

$$P(\lambda z_1 + (1-\lambda)z_2) \leq \lambda \inf_{x_1} \{f(x_1) : x_1 \in X, g(x_1) - z_1 \in K\}$$

$$+ (1-\lambda) \inf_{x_2} \{f(x_2) : x_2 \in X, g(x_2) - z_2 \in K\}$$

$$= \lambda P(z_1) + (1-\lambda)P(z_2).$$

Hence P is convex.

The following lemma gives a monotonicity property of P. If K is the nonnegative orthant, then it implies P is a nondecreasing function. If K is the nonpositive orthant, then it implies P is a nonincreasing function.

8.1.7 Lemma. Consider the perturbation function P of Definition 8.1.4. If $z_1 - z_2 \in K$, then $P(z_1) \geq P(z_2)$.

Proof: Let $A_1 = \{x \in X : g(x) - z_1 \in K\}$ and $A_2 = \{x \in X : g(x) - z_2 \in K\}$. Since K is a convex cone, then under the assumption $z_1 - z_2 \in K$, it is immediate that $x \in A_1$ implies $x \in A_2$, i.e., $A_1 \subset A_2$. Hence $P(z_1) = \inf \{f(x) : x \in A_1\} \geq \inf \{f(x) : x \in A_2\} = P(z_2)$.

8.2. The Strong Duality Theorem

We will now prove the main duality theorem which states that the optimal objectives of the primal and dual problems are equal. In this section both the inequality and equality constraints are handled explicitly. More specifically the primal and dual problems are:

Primal

Minimize $f(x)$

Subject to $x \in X$

$g(x) \in C$

$h(x) = 0$

Dual

Maximize $\theta(u,v)$

Subject to $u \in C^*$
v unrestricted

where $\theta(u,v) = \inf_{x \in X}\{f(x) + \langle u,g(x)\rangle + \langle v,h(x)\rangle\}$

Note that the above formulation is the same as that of Section 8.1 where $K = C \times \{0\}$ and g is replaced by (g,h).

8.2.1 Theorem. Let X be a convex set in E_n and C be a closed convex cone in E_m. Let $f : E_n \to E_1$ be convex and $g : E_n \to E_m$ be C-convex. Let $h : E_n \to E_k$ be linear. Consider the primal and dual problems defined above. Let $\inf\{f(x) : x \in X, g(x) \in C, h(x) = 0\} = \gamma$ be finite. Also suppose that

(i) $0 \in \operatorname{int} h(X)$ where $h(X) = \{h(x) : x \in X\}$

(ii) There exists an $x \in X$ such that $g(x) \in \operatorname{int} C$ and $h(x) = 0$.

Then

$$\gamma = \inf\{f(x) : x \in X, g(x) \in C, h(x) = 0\} = \max\{\theta(u,v) : u \in C^*, v \in E_k\}$$
$$= \theta(\bar{u},\bar{v})$$

for some $\bar{u} \in C^*$ and $\bar{v} \in E_k$. Furthermore, if the inf on the left hand side is achieved by some $\bar{x}$, then $\langle\bar{u},g(\bar{x})\rangle = 0$.

Proof: Consider the system

$$f(x) - \gamma < 0, \quad g(x) \in C, \quad h(x) = 0, \quad x \in X.$$

By definition of γ, the system has no solution. From Theorem 7.1.2 of Chapter 7, there exists a nonzero vector $(\bar{u}_0, \bar{u}, \bar{v})$ with $\bar{u}_0 \geq 0$, $\bar{u} \in C^*$ and $\bar{v} \in E_k$ such that

$$\bar{u}_0\,(f(x) - \gamma) + \langle\bar{u},g(x)\rangle + \langle\bar{v},h(x)\rangle \geq 0 \text{ for all } x \in X.$$

Using an argument similar to that in the proof of Lemma 7.1.7, if $\bar{u}_0 = 0$, then $(\bar{u}_0, \bar{u}, \bar{v}) = 0$ which is impossible. Hence $\bar{u}_0 > 0$, and without loss of generality we let $\bar{u}_0 = 1$. Hence we get

$$f(x) + \langle\bar{u},g(x)\rangle + \langle\bar{v},h(x)\rangle \geq \gamma \text{ for all } x \in X.$$

This shows that

$$\theta(\bar{u},\bar{v}) = \inf_{x \in X}\{f(x) + \langle\bar{u},g(x)\rangle + \langle\bar{v},h(x)\rangle\} \geq \gamma .$$

But from Theorem 8.1.3, $\sup\{\theta(u,v) : u \in C^*\} \leq \gamma$. Hence $\theta(\bar{u},\bar{v}) = \gamma$, and the first part of the theorem is at hand.

To prove the second part, suppose that $\bar{x}$ solves the primal problem, i.e. $\inf\{f(x) : x \in X,\ g(x) \in C,\ h(x) = 0\} = f(\bar{x})$. Therefore, we get

$$f(\bar{x}) = \theta(\bar{u},\bar{v}) = \inf_{x \in X} \{f(x) + \langle \bar{u}, g(x)\rangle + \langle \bar{v}, h(x)\rangle\}$$

$$\leq f(\bar{x}) + \langle \bar{u}, g(\bar{x})\rangle$$

since $h(\bar{x}) = 0$. This implies $\langle \bar{u}, g(\bar{x})\rangle \geq 0$. But, on the other hand, $\bar{u} \in C^*$ and $g(\bar{x}) \in C$, implies $\langle \bar{u}, g(\bar{x})\rangle \leq 0$. Hence $\langle \bar{u}, g(\bar{x})\rangle = 0$ and the proof is complete.

The above theorem shows, under appropriate convexity assumptions, that the optimal objectives of the primal and dual problems are equal. Hence the primal problem can be indirectly solved by solving the dual problem. Also note that if $(\bar{u},\bar{v})$ is an optimal dual solution, then $\bar{x}$, the optimal primal solution, is an optimal solution of the problem

$$\min_{x \in X} \{f(x) + \langle \bar{u}, g(x)\rangle + \langle \bar{v}, h(x)\rangle\}$$

It may be noted that the qualification on the constraints adopted in the theorem, which is a generalization of the Uzawa qualification of Chapter 6, guarantees that the lagrangian multiplier associated with the objective function is positive, and that the primal and dual objectives are equal. We will show, later in the section, that any constraint qualification that insures that the Lagrangian multiplier of the objective function is positive in the saddle point optimality conditions will also insure that a duality gap (difference between the optimal primal and dual objectives) will not arise. In other words, duality gaps can arise only when the multiplier of the objective function is zero.

Theorem 8.2.2 below characterizes the optimal solution of the dual problem in [illegible]ms of subgradients of the perturbation function at $z = 0$.

[illegible] Theorem. Let X be a convex set in E_n and K be a closed convex cone in E_m. [illegible] $E_n \to E_1$ be convex and $g : E_n \to E_m$ be K-convex. Now consider the primal and [illegible]ems defined 8.1.1 and the perturbation function defined in 8.1.4. Let [illegible] $f(x),\ x \in X,\ g(x) \in K\} = \gamma$ be finite. Then $\bar{u}$ is an optimum solution to [illegible]em if and only if $-\bar{u}$ is a subgradient of P at 0.

Proof: First assume that $-\bar{u}$ is a subgradient of P at 0, i.e., $P(z) \geq P(0) - \langle\bar{u},z\rangle$ for all z. If $-z \in K$, then $0-z \in k$, and by Lemma 8.1.7 $P(0) \geq P(z)$. We then have $\langle\bar{u},z\rangle \geq P(0) - P(z) \geq 0$ for all $-z \in K$ which implies that $\bar{u} \in K^*$. We have thus shown that $\bar{u}$ is feasible to the dual problem. Let $x \in X$. Then $P(g(x)) = \inf \{f(y) : y \in X, g(y) - g(x) \in K\} \leq f(x)$. Therefore $f(x) \geq P(g(x)) \geq P(0) - \langle\bar{u},g(x)\rangle$ since $-\bar{u}$ is a subgradient of P at 0. Since this is true for each $x \in X$ then $\theta(\bar{u}) = \inf_X \{f(x) + \langle\bar{u},g(x)\rangle\} \geq P(0) = \gamma$. This shows that $\bar{u}$ is an optimal dual solution.

To prove the converse, suppose that $\bar{u}$ solves the dual problem, i.e., $P(0) = \theta(\bar{u}) = \inf \{f(x) + \langle\bar{u},g(x)\rangle : x \in X\}$ where $\bar{u} \in K^*$. Fix any $z \in Em$ and consider the set $\{x \in X : g(x) - z \in K\}$. Since $\bar{u} \in K^*$, then $\langle\bar{u},g(x)\rangle \leq \langle\bar{u},z\rangle$ and therefore we get $f(x) + \langle\bar{u},z\rangle \geq f(x) + \langle\bar{u},g(x)\rangle \geq \theta(\bar{u}) = P(0)$. Since this is true for each $x \in X$ with $g(x) - z \in K$, $\inf \{f(x) : x \in X, g(x) - z \in K\} + \langle\bar{u},z\rangle \geq P(0)$, or $P(z) \geq P(0) - \langle\bar{u},z\rangle$. But since this is true for each z, then $-\bar{u}$ is a subgradient of P at 0 and the proof is complete.

It may be recalled that a function f is said to be subdifferentiable at a point if the set of subgradients of f at the point is nonempty. Thus, Theorem 8.2.2 above asserts that the dual problem has an optimal solution if and only if P is subdifferentiable at 0, where the negative of any subgradient is an optimal dual solution.

We now present a theorem dealing with the case when the primal constraints are inconsistent.

8.2.3 Theorem. Consider the primal and dual problems defined in 8.1.1 and suppose that X is contex and g in K-convex. Suppose that $\{x \in X : g(x) \in K\} = \emptyset$ and assume that $\sup_{u \in K^*} \theta(u)$ is finite. Then $0 \in c\ell\, Z$, where $Z = \{z : \{x \in X : g(x) - z \in K\} \neq \emptyset\}$.

Proof: Suppose that $0 \notin c\ell\, Z$. Z is convex by noting convexity of X and K-convexity of g. Then there is a hyperplane that strictly separates 0 and $c\ell\, Z$ (and hence Z), i.e., there exists a nonzero q and an ϵ such that $\langle q,z\rangle \geq \epsilon > 0$ for all $z \in Z$. We note that $q \in K^*$. To show this let $\xi \in K$ and fix an $x \in X$. Now let $z_\lambda = g(x) - \lambda\xi$, then $z_\lambda \in Z$ since $g(x) - z_\lambda \in K$ for each $\lambda > 0$. Now $\langle q,g(x)\rangle - \lambda\langle q,\xi\rangle > 0$ for each $\lambda > 0$ implies $\langle q,\xi\rangle \leq 0$ since otherwise by choosing λ sufficiently large we violate the above inequality. So $q \in K^*$. Now for each $x \in X$ we have $g(x) \in Z$. Therefore $\langle q,g(x)\rangle \geq \epsilon$ for each $x \in X$. Hence we get $\inf \{\langle q,g(x)\rangle :$

$x \in X\} \geq \epsilon > 0$. Since $\sup_{u \in K^*} \theta(u)$ is finite, there exists a $u \in K^*$ such that $\theta(u)$ is finite. Consider $u + \alpha q$ for $\alpha > 0$. Note that $u + \alpha q \in K^*$.

Furthermore

$$\theta(u + \alpha q) = \inf_{x \in X} \{f(x) + \langle(u + \alpha q), g(x)\rangle\} \geq \inf_{x \in X} \{f(x) + \langle u, g(x)\rangle\}$$

$$+ \alpha \inf_{x \in X} \{\langle q, g(x)\rangle\}$$

$$\geq \theta(u) + \alpha \epsilon$$

By letting $\alpha \to \infty$ we see that $\theta(u + \alpha q) \to \infty$ which violates our assumption.

Corollary. Suppose that X is a compact convex set, f is continuous, and g is a continuous K-convex function. If $[x \in X : g(x) \in K\} = \emptyset$ then $\inf\{f(x) : x \in X, g(x) \in K\} = \sup_{u \in K^*} \theta(u) = \infty$.

Proof: If $\inf\{f(x) : x \in X, g(x) \in K\} = \emptyset$ then $\inf\{f(x) : x \in X, g(x) \in K\} = + \infty$ by convention. The result follows by applying the theorem and noting that $0 \notin Z$ and X compact insures that z is closed.

It may be noted that the dual problem of linear programming readily follows from the above formulation. Furthermore, for a general convex program: min $f(x)$ s.t. $g(x) \leq 0$ where f and g are convex differentiable functions, the dual can be written as

$$\text{Max } \{f(x) + \langle u, g(x)\rangle\}$$

subject to

$$u \geq 0, \quad \nabla f(x) + \langle \nabla g(x), u\rangle = 0$$

which is Wolfe's duality formulation.

CHAPTER 9

CONJUGATE DUALITY

In recent years conjugate function theory has been used frequently in nonlinear programming, particularly in the context of duality. Even though the initial results were restricted to convex functions, the approach has been found quite powerful for nonconvex functions also. In this chapter we will consider this more general problem where the functions involved may be nonconvex.

In discussing conjugate functions it is convenient to consider "closed" functions, i.e., functions whose epigraphs are closed. We will discuss below how a function can be "closed" if it is not already so. We will observe later that this operation will simplify discussion in subsequent sections without loss of generality.

9.1. Closure of a Function

It may be noted that the epigraph of a function is not necessarily closed even if the function is convex. By redefining the values of the function at appropriate points, we can deal with a function whose epigraph is closed. This leads to the definition of a lower closure of a function f, denoted by $\underline{f}$. Later, we will define an upper closure $\bar{f}$ which will be useful in considering maximization problems. It may be noted the concept of closing a function was used earlier in Chapter 4.

9.1.1 Definition. Let S be a nonempty set in E_n and let $f : S \to E_1$. The lower closure of f, denoted by $\underline{f}$, is given by

$$\underline{f}(x) = \underline{\lim}_{z \to x} f(z)$$

for each $x \in C\ell\, S$. If $f = \underline{f}$ then f is said to be lower closed on S.

We note that a convex function f is continuous on ri(dom f), and hence f and $\underline{f}$ are identical on ri(dom f). Hence, when f is convex, $\underline{f}$ is obtained by redefining the values of the function at points in the relative boundary.

The following theorem shows that indeed $E_{\underline{f}}$ is a closed set; namely, the closure of E_f. Notice that no convexity of f is assumed.

9.1.2 Theorem. Let S be a nonempty set in E_n and $f : S \to E_1$.

Then

$$E_{\underline{f}} = C\ell\, E_f$$

where $E_{\underline{f}}$ and E_f are the epigraphs of $\underline{f}$ and f respectively.

Proof: We will first show that $E_{\underline{f}} \subset C\ell\, E_f$. Let $(x,y) \in E_{\underline{f}}$, i.e., $y \geq \underline{f}(x)$ and $x \in C\ell\, S$. This means that $y \geq \underline{f}(x) = \underset{z \to x}{\ell im} f(z)$, which implies that there exists a sequence $\{z_k\}$ in S converging to x such that $\underline{f}(x) = \underset{k \to \infty}{\ell im} f(z_k)$. This, in turn, implies that there exists a sequence of scalars $\{\epsilon_k\}$ converging to zero such that $\underline{f}(x) + \epsilon_k = f(z_k)$ for all k. But $y \geq \underline{f}(x)$. Hence $y + \epsilon_k \geq f(z_k)$ for all k, and $\epsilon_k \to 0$, $z_k \to x$. Letting $y_k = y + \epsilon_k$, then we have constructed a sequence $(z_k, y_k) \in E_f$ with limit (x,y) which implies that $(x,y) \in C\ell\, E_f$. This shows that $E_{\underline{f}} \subset C\ell\, E_f$. To show the converse inclusion let $(x,y) \in C\ell\, E_f$, i.e., $(x,y) = \lim\, (x_k, y_k)$ with $y_k \geq f(x_k)$, $x_k \in S$. Then $y = \underset{k \to \infty}{\lim} y_k \geq \underset{x_k \to x}{\underline{\lim}} f(x_k) \geq \underset{z \to x}{\underline{\lim}} f(z) = \underline{f}(x)$, i.e., $(x,y) \in E_{\underline{f}}$. This completes the proof.

Note that a discrete function f always satisfies the above definition if we let the sequence $\{z_k\} = x$ for all k. That is, such a function is always closed. The following theorem gives a characterization of closed functions in terms of level sets.

9.1.3 Theorem. Let S be a nonempty set in E_n and let $f : S \to E_1$. Then $f = \underline{f}$ if and only if the set

$$\Lambda_\alpha = \{x : x \in S\ ,\ f(x) > \alpha\}$$

is open relative to S for each real α.

Proof: First we will show that if $f = \underline{f}$, then Λ_α is open relative to S. Choose an $\alpha \in E_1$ and suppose $\Lambda_\alpha \neq \emptyset$. Let $x_o \in \Lambda_\alpha$. Then there exists an $\epsilon > 0$ such that $f(x_o) - \epsilon > \alpha$. Now since $f = \underline{f}$, then $f(x_o) = \underline{f}(x_o) = \underset{z \to x_o}{\underline{\lim}} f(z)$, i.e., there exists a neighborhood $N_\delta(x_o)$ such that $f(z) \geq f(x_o) - \epsilon > \alpha$ for each $z \in N_\delta(x_o) \cap S$. That is, $z \in N_\delta(x_o) \cap S$ implies $z \in \Lambda_\alpha$, and hence Λ_α is open relative to S.

To prove the converse, suppose Λ_α is open relative to S for each real α. Let $x_o \in S$ and choose $\epsilon > 0$. Let $\hat{\alpha} = f(x_o) - \epsilon$. Then clearly $x_o \in \Lambda_{\hat{\alpha}}$. Since $\Lambda_{\hat{\alpha}}$ is open relative to S, there exists a neighborhood $N_\delta(x_o)$ such that $f(z) > \hat{\alpha} = f(x_o) - \epsilon$ for each $z \in N(x_o) \cap S$. Hence $\underset{z \to x_o}{\underline{\lim}} f(z) + \epsilon \geq f(x_o)$. Since this is true for each $\epsilon > 0$ then $\underset{z \to x_o}{\underline{\lim}} f(z) \geq f(x_o)$. But $\underset{z \to x_o}{\underline{\lim}} f(x) \leq f(x_o)$ and so $\underline{f}(x_o) = f(x_o)$.

The following lemma states that the sum of two lower closed functions is lower closed. The proof follows immediately from the Definition 9.1.1.

9.1.4 Lemma. Let S be a nonempty set in E_n. Let $f_1 : S \to E_1$ and $f_2 : S \to E_1$ be two lower closed functions. Then $f = f_1 + f_2$ is lower closed on S.

The following theorem gives another important characterization of the closure of a convex function f.

9.1.5 Theorem. Let S be a nonempty convex so on E_n and let $f : S \to E_1$ be convex. Let $D = \{(\xi,\mu) : f(x) \geq \langle x,\xi\rangle - \mu$ for each $x \in S\}$. Then $\underline{f}(x) = \sup \{\langle x,\xi\rangle - \mu : (\xi,\mu) \in D\}$.

Proof: We will first show that $\underline{f}(x) \geq \sup \{\langle x,\xi\rangle - \mu : (\xi,\mu) \in D\}$. We will use a contradiction argument as follows. Suppose for some $(\xi_o,\mu_o) \in D$ we have $\underline{f}(x) < \langle x,\xi_o\rangle - \mu_o$ and hence $f(x) = \langle x,\xi_o\rangle - \mu_o - \epsilon$ for some $\epsilon > 0$. By definition $\underline{f}(x) = \underset{z \to x}{\underline{\lim}} f(z)$ and hence there is a sequence $\{z_k\}$ in S with $z_k \to x$ and $\underline{f}(x) = \underset{k \to \infty}{\lim} f(z_k)$. Therefore, $\underline{f}(x) = \lim_{k\to\infty} f(z_k) = \langle x,\xi_o\rangle - \mu_o - \epsilon = \langle z_k,\xi_o\rangle - \mu_o - \epsilon + \langle x - z_k,\xi_o\rangle$ for all k. By definition of $\underline{f}$, however, there must exist a sequence of scalars ϵ_k such that $\epsilon_k \to 0$ and $f(z_k) + \epsilon_k \underline{f}(x)$. So we get $f(z_k) = \langle z_k,\xi_o\rangle - \mu_o - (\epsilon + \epsilon_k) + \langle x - z_k,\xi_o\rangle$. Since $z_k \to x$ and $\epsilon_k \to 0$ then there is an integer N such that $-\epsilon - \epsilon_N + \langle x - z_N,\xi_o\rangle < 0$ and so we get $f(z_N) < \langle z_N,\xi_o\rangle - \mu_o$ which contradicts the assumption that $(\xi_o,\mu_o) \in D$ and $z_N \in S$. This then shows that $\underline{f}(x) \geq \sup \{\langle x,\xi\rangle - \mu : (\xi,\mu) \in D\}$. We will now show that $\underline{f}(x) \leq \sup \{\langle x,\xi\rangle - \mu : (\xi,\mu) \in D\}$. Let $x_o \in \text{ri } S$ and let $x_k = \frac{1}{k} x_o + \left(1 - \frac{1}{k}\right) x \in \text{ri } S$ for each $k \geq 1$. By Theorem 4.1.5 there exists a "nonvertical" supporting hyperplane of E_f passing through $(x_k,f(x_k))$.

In other words there exists $(\xi_k,-1)$ such that $f(x) \geq f(x_k) + \langle x-x_k,\xi_k\rangle$ for all $x \in S$. Letting $\mu_k = \langle x_k,\xi_k\rangle - f(x_k)$ then $(\xi_k,\mu_k) \in D$. Therefore

$$\begin{aligned} f(x_k) &= \langle x_k,\xi_k\rangle - \mu_k \leq \sup \{\langle x_k,\xi\rangle - \mu : (\xi,\mu) \in D\} \\ &= \sup \{\langle \tfrac{1}{k} x_o + \left(1 - \tfrac{1}{k}\right)x,\xi\rangle - \mu :: (\xi,\mu) \in D\} \\ &\leq \frac{1}{k} \sup \{\langle x_o,\xi\rangle - \mu : (\xi,\mu) \in D\} + \left(1 - \frac{1}{k}\right) \sup\{\langle x,\xi\rangle - \mu : (\xi,\mu) \in D\} \end{aligned}$$

Note that $\sup \{\langle x_o, \xi\rangle - \mu : (\xi,\mu) \in D\} \le \underline{f}(x_o) = f(x_o)$ since $x_o \in$ ri S and f is continuous on ri S. Therefore

$$f(x_k) \le \frac{1}{k} \underline{f}(x_o) + \left(1 - \frac{1}{k}\right) \sup \{\langle x, \xi\rangle - \mu : (\xi,\mu) \in D\}$$

which implies that $\overline{\lim_{k\to\infty}} f(x_k) \le \sup\{\langle x, \xi\rangle - \mu : (\xi,\mu) \in D)$. Since $x_k \to x$, the above inequality in turn implies that $\underline{f}(x) \le \sup\{\langle x,\xi\rangle - \mu : (\xi,\mu) \in D\}$ and the proof is complete.

We note in the above theorem that the set D corresponds to all hyperplanes of the form $\{(x,y) : \langle (x,y) , (\xi,-1)\rangle = \mu\}$ and which is dominated by the graph of the function f, i.e., each of these hyperplanes divide the Euclidean space into two halfspaces with E_f contained in one of these halfspaces. Now we may interpret $\underline{f}(x)$ to be the supremum of the value of y over (x,y) belonging to these hyperplanes. Of course, if $x \in$ ri S then we actually have a supporting hyperplane of E_f at $(x,f(x))$ and hence the supremum is achieved at $y = f(x)$, i.e. $\underline{f}(x) = f(x)$.

The notion of upper closure which is useful for maximization problems is given below. Results si ilar to Theorems 9.1.2, 9.1.3, 9.1.5 and Lemma 9.1.4 can be drawn for the upper closure.

9.1.6 Definition. Let S be a nonempty set in E_n and let $g : S \to E_1$. The upper closure of g, denoted by $\bar{g}$, is given by

$$\bar{g}(x) = \overline{\lim_{z \to x}} g(z) \quad \text{for } x \in C\ell\, S$$

It $g = \bar{\xi}$ then g is called upper closed.

9.2. Conjugate Functions

In this section we will define and examine some basic properties of conjugate functions. We start with the definition of convex conjugate functions below.

9.2.1 Definition. Let S be a nonempty set in E_n and let $f : S \to E_1$.

Let $S' = \{\xi : \sup_{x \in S} \{\langle \xi,x\rangle - f(x) < \infty\}$. Then $f' : S' \to E_1$ defined by $f'(\xi) = \sup_{\xi \in S'} \{\langle \xi,x\rangle - f(x)\}$ is said to be the convex conjugate function of f. We will simply say that f' is the conjugate of f as long as no confusion arises.

The following lemma shows that f' is a convex function and S' is a convex set.

Note that this result holds true even in the absence of convexity of f or S.

9.2.2 Lemma. S′ is a convex set and f′ is a convex function on S′.

Proof: Let ξ_1 and $\xi_2 \in S'$ and consider $\lambda\xi_1 + (1 - \lambda)\xi_2$ with $\lambda \in (0,1)$.

$$\sup_{x \in S} \{\langle\lambda\xi_1 + (1 - \lambda)\xi_2, x\rangle - f(x)\} = \sup_{x \in S} \{\lambda(\langle\xi_1, x\rangle - f(x)) + (1 - \lambda)(\langle\xi_2, x\rangle - f(x)\}$$

$\le \lambda \sup_{x \in S} \{\langle\xi_1, x\rangle - f(x)\} + (1 - \lambda) \sup_{x \in S} \{\langle\xi_2, x\rangle - f(x)\}$. But since $\xi_1, \xi_2 \in S'$ then the right hand side is $< \infty$ and hence we have established convexity of S′. Noting that the left hand side of the above inequality is $f'(\lambda\xi_1 + (1 - \lambda)\xi_2)$ and the right hand side is $\lambda f'(\xi_1) + (1 - \lambda)f'(\xi_2)$ then we also have established convexity of f′ on S′. This completes the proof.

We note the following interpretation of the conjugate function f′. For a given $\xi \in S'$, $\langle\xi, x\rangle - y = 0$ is an equation of a hyperplane in E_{n+1} through the origin. To find $f'(\xi)$ we must find $\sup_{x \in S} \{\langle\xi, x\rangle - f(x)\}$. In Figure 9.1 we want to maximize $\langle\xi, x\rangle - f(x)$ or equivalently minimize the $f(x) - \langle\xi, x\rangle$, i.e., we move the hyperplane $\{(x,y) : \langle\xi, x\rangle - y = 0\}$ vertically until it supports the epigraph of f. The intercept of this hyperplane on the vertical axis would then give us $-f'(\xi)$. In other words we may think of $-f'(\xi)$ as the intercept on the vertical axis of the hyperplane that supports E_f with normal vector $(\xi, -1)$. Of course, the case when $\xi \in S'$ (i.e., sup is finite) corresponds to a nonvertical supporting hyperplane, whereas the case when $\xi \notin S'$ corresponds to a <u>vertical</u> supporting hyperplane.

An important special case that when $f(x) = 0$ for all $x \in S$. Here $f'(g) = \sup_{x \in S} \langle\xi, x\rangle$. This function is the support function of the set S. $f'(\xi)$ corresponds to the hyperplane having normal ξ and supporting the set S. In other words, we may interpret supporting hyperplanes to sets via the notion of conjugacy. In this special case S′ is a convex cone since $\sup_{x \in S} \langle\xi, x\rangle < \infty$ then $\sup_{x \in S} \langle\lambda\xi, x\rangle < \infty$ for any $\lambda > 0$, i.e., S′ is a cone. Convexity of S′ was established earlier. Also, f′ is a positively homogeneous convex function, i.e., f′ is convex and $\lambda f'(\xi) = f'(\lambda\xi)$ for all $\lambda > 0$. That is, f′ satisfies the following properties:

(i) $f'(\xi_1 + \xi_2) \le f'(\xi_1) + f'(\xi_2)$

(ii) $f'(\lambda\xi) = \lambda f'(\xi)$ for all $\lambda > 0$.

There is an interesting relationship between the closure of a convex function f

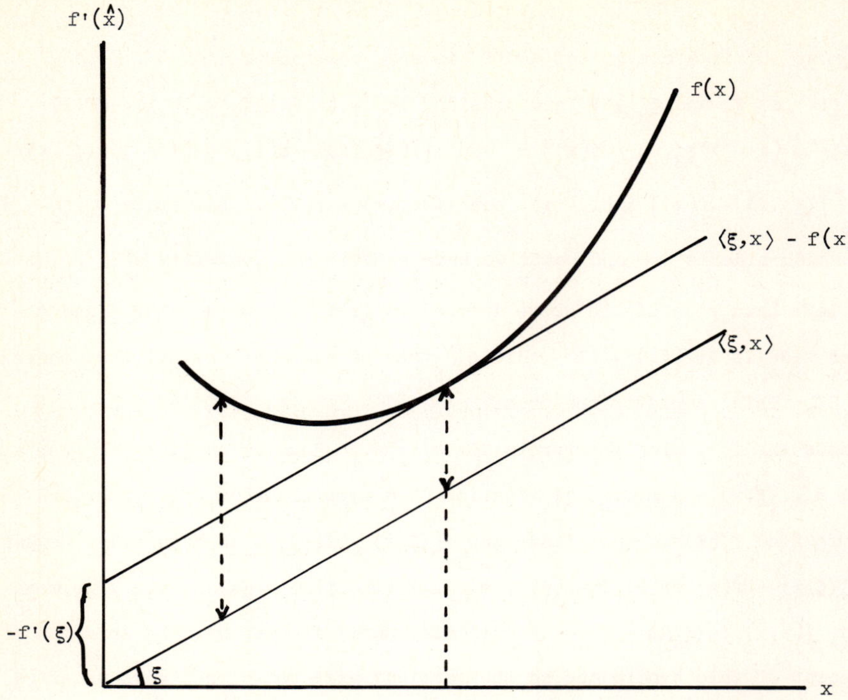

Figure 9.1. Interpretation of Conjugate Convex Functions.

and the conjugate of the conjugate function of f. This is made precise in the following theorem.

9.2.3 <u>Theorem</u>. Let S be a nonempty convex set in E_n and let $f : S \to E_1$ be a convex function. Let $D = \{(\xi,\mu) : f(x) \geq \langle x,\xi\rangle - \mu \text{ for each } x \in S\}$ and let $S_o = \{x \in C\ell\, S : \sup \{\langle x,\xi\rangle - \mu : (\xi,\mu) \in D\} < \infty\}$. Then $S_o = S''$ and $\underline{f} = f''$.

<u>Proof</u>: Let $\bar{x} \in S_o$. Noting Theorem 9.1.5 we get

$$\infty > \underline{f}(\bar{x}) = \sup \{\langle \bar{x},\xi\rangle - \mu : (\xi,\mu) \in D\}$$

Now let $\xi \in S'$. By definition of f' it follows that $f(x) \geq \langle \xi,x\rangle - f'(\xi)$ for all $x \in S$. This shows that $(\xi, f'(\xi)) \in D$ and hence $\infty > \underline{f}(\bar{x}) \geq \langle \bar{x},\xi\rangle - f'(\xi)$. Since this is true for each $\xi \in S'$ then $\infty > \underline{f}(\bar{x}) \geq \sup_{\xi \in S'} \{\langle \bar{x},\xi\rangle - f'(\xi)\} = f''(\bar{x})$.

This also shows that $\bar{x} \in S''$ and hence $S_o \subset S''$. Now let $\bar{x} \in S''$. Then $\infty > f''(\bar{x}) = \sup_{\xi \in S'} \{\langle \xi,\bar{x}\rangle - f'(\xi)\}$. Let $(\xi,\mu) \in D$, i.e., $\mu \geq \langle \xi,x\rangle - f(x)$ for all $x \in S$. By definition $f'(\xi) = \sup_{x \in S} \{\langle \xi,x\rangle - f(x)\}$ and it is obvious that $\sup_{x \in S} [\langle \xi,x\rangle\} < \infty$ because otherwise we cannot find a μ with $\mu \geq \langle \xi,x\rangle - f(x)$ for each $x \in S$. Hence $\xi \in S'$. So we get $\mu \geq f'(\xi)$ for each $(\xi,\mu) \in D$. But $f''(\bar{x}) \geq \langle \xi,\bar{x}\rangle - f'(\xi)$ for each $\xi \in S'$ and hence $f''(\bar{x}) \geq \langle \xi,\bar{x}\rangle - \mu$ for each $(\xi,\mu) \in D$. So $\infty > f''(\bar{x}) \geq \sup \{\langle \xi,\bar{x}\rangle - \mu : (\xi,\mu) \in D\}$. This also shows that $\bar{x} \in S_o$ and hence $S'' \in S_o$. This completes the proof.

In the above theorem we have assumed S to be a convex set and f to be a convex function. We can get some interesting and useful results even if we relax the convexity assumption. For the sake of clarity and convenience we will assume below that S is compact and f is closed. A little reflection will show that these "closedness" properties can be relaxed to obtain further generalizations. Before discussing the properties of conjugate functions associated with nonconvex functions, we will need the following definition.

9.2.4 <u>Definition</u>. Let S be a nonempty compact set in E_n and let $f : S \to E_1$ be closed. Then the <u>convex envelope</u> of f, denoted by f_{co}, is defined by

$$f_{co}(x) = \inf \{y : (x,y) \in H(E_f)\}$$

where $H(E_f)$ is a convex hull of the epigraph of f. The domain of f_{co} is defined by

$$S_o = \{x : (x,y) \in H(E_f) \quad \text{for some } y\} .$$

The following Lemmas show that for a closed function f, S_o and the domain of the conjugate of the conjugate function of f are both equal to the convex hull of S.

9.2.5 <u>Lemma</u>. Let S be a nonempty compact set in E_n and let $f : S \to E_1$ be closed. Then the set S_o of Definition 9.2.4 is convex and $S_o = H(S)$, the convex hull of S.

<u>Proof</u>: Let $x \in S$. Then $(x,f(x)) \in E_f$ and by definition of S_o, it is clear that $x \in S_o$. Hence $S \subset S_o$. It is readily verified that S_o is a convex set, and hence $H(S) \subset S_o$. To show the reverse inclusion, let $x \in S_o$. Then, by definition there exists a y such that $(x,y) \in H(E_f)$. Since $H(E_f) \subset E_{n+1}$, then by Theorem 2.2.4, (x,y) can be expressed as

$$(x,y) = \sum_{i=1}^{n+2} \lambda_i(x_i,y_i) , \sum_{i=1}^{n+2} \lambda_i = 1, \lambda_i \geq 0 \quad i = 1,\ldots,n+2$$

when $(x_i,y_i) \in E_f$. This implies x can be written as

$$x = \sum_{i=1}^{n+2} \lambda_i x_i, \sum_{i=1}^{n+2} \lambda_i = 1, \lambda_i \geq 0 \quad i = 1,\ldots,n+2$$

with $x_i \in S$ for each $i = 1,2,\ldots,n+2$, $x \in H(S)$ or $S_o \subset H(S)$. Hence $S_o = H(S)$.

9.2.6 Lemma. Let S be a nonempty compact set in E_n and let $f : S \to E_1$ be closed. Then $S'' = H(S)$ where

$$S'' = \{x : \sup_{\xi \in S'} \{\langle x,\xi\rangle - f'(\xi)\} < \infty\} .$$

Proof: Consider $x \in S$. Then $\langle \xi,x\rangle - f(x) \leq f'(\xi)$ for each $\xi \in S'$. Hence $\langle \xi,x\rangle - f'(\xi) \leq f(x)$ for each $\xi \in S'$, i.e., $\sup_{\xi \in S'} \langle \xi,x\rangle - f'(\xi) \leq f(x) < \infty$ since f is finite on S. Hence $x \in S''$, i.e., $S \subset S''$. But from Lemma 9.2.2, S'' is convex. Hence $H(S) \subset S''$.

To complete the proof we need to show the reverse inclusion, i.e., $S'' \subset H(S)$. Assume there is a point $x_o \in S''$ but not in H(S). From Theorem 2.4.4 and noting that S is compact, there exists a unique minimizing point $x^* \in H(S)$ such that

$$\inf_{x \in S} \{\langle x^* - x_o, x - x_o\rangle\} \geq \inf_{x \in H(S)} \{\langle x^* - x_o, x - x_o\rangle\} = \gamma > 0.$$

Since S is compact then $\sup_{x \in S} \{\langle \lambda(x_o - x^*), x\rangle - f(x)\} < \infty$ for any $\lambda \in E_1$ and so $\lambda(x_o - x^*) \in S'$. Since $x_o \in S''$ it then follows that

$$f''(x_o) \geq \langle x_o,\xi\rangle - f'(\xi) \quad \text{for each } \xi \in S' .$$

Again, by definition, for $\xi \in S'$,

$$f'(\xi) = \sup_{x \in S} \{\langle \xi,x\rangle - f(x))\}$$

$$\leq \sup_{x \in S} \{\langle \xi,x\rangle + \sup_{x \in S} \{-f(x)\} .$$

Substituting in the above inequality we get:

$$f''(x_o) \geq \langle x_o, \xi\rangle - \sup_{x \in S} \langle \xi, x\rangle - \sup_{x \in S} (-f(x))$$

$$= \inf_{x \in S} \langle \xi, x_o - x\rangle - \sup_{x \in S} (-f(x))$$

Letting ξ be $\lambda(x_o - x^*)$ we get:

$$f''(x_o) \geq \lambda \inf_{x \in S} \langle x_o - x^*, x_o - x\rangle - \sup_{x \in S} (-f(x))$$

$$= \lambda\gamma - \sup_{x \in S} (-f(x))$$

Since $\gamma > 0$, letting λ approach $+\infty$ we conclude that $f''(x_o) = \infty$ contradicting the assumption that $x_o \in S''$. Therefore $x_o \in H(S)$ and the proof is complete.

9.2.7 Lemma. Let S be a nonempty compact set in E_n and let $f : S \to E_1$ be closed. Then

$$f_{co}(x) = f''(x) \quad \text{for each } x \in H(S) .$$

Proof: By definition $f'(\xi) = \sup_{x \in S} \{\langle x,\xi\rangle - f(x)\}$. Hence

$$\langle x,\xi\rangle - f'(\xi) \leq f(x) \quad \text{for each } x \in S .$$

Hence,

$$f''(x) = \sup_{\xi \in S'} \{\langle x,\xi\rangle - f'(\xi)\} \leq f(x) \quad \text{for each } x \in S.$$

This implies $E_{f''} \supset E_f$. But $H(E_f)$ is the smallest convex set containing E_f. Hence $E_{f''} \supset H(E_f)$. This shows that $f''(x) \leq f_{co}(x)$ for each $x \in H(S)$.

To prove that $f''(x) \geq f_{co}(x)$, we need to show that $E_{f''} \subset E_{f_{co}} = H(E_f)$. Suppose that $(x_o,y_o) \in E_{f''}$. Therefore $x_o \in S''$ and $f(x_o) \leq y_o$. But from Lemmas 9.2.5 and 9.2.6, $S'' = H(S) = S_o$. Hence $x_o \in S_o$ and $f(x_o) \leq y_o$, i.e., $(x_o,y_o) \in E_{f_{co}}$. Therefore $E_{f''} \subset E_{f_{co}}$ and thus $f''(x) \geq f_{co}(x)$. This completes the proof.

Definition 9.2.8 below presents concave conjugate functions. Results similar to those of convex conjugate functions can be drawn for the concave case.

9.2.8 Definition. Let $g : S \to E_1$ where S is in E_n. The concave conjugate function g' is defined by

$$g'(\xi) = \inf_{x \in S} \{\langle x, \xi \rangle - g(x)\} \quad \text{for all } \xi \in S'$$

where

$$S' = \{\xi : \inf_{x \in S} \{\langle x, \xi \rangle - g(x)\} > -\infty\} .$$

9.3. Main Duality Theorem

In this section we will develop an important duality theorem using conjugate functions. Consider the following primal problem:

$$P : \text{Minimize}_{x \in S_1 \cap S_2} \{\alpha_1(x) - \alpha(x)\} ,$$

where α_1 and α_2 are functions defined on subsets S_1 and S_2 in E_n respectively.

It may be noted that the above problem can be represented in the more familiar form of a nonlinear programming problem by properly choosing the functions α_1 and α_2 and the sets S_1 and S_2. We will do this in detail later.

Now we will consider the following dual program:

$$D : \text{Maximize}_{\xi \in S_1' \cap S_2'} \{\alpha_2'(\xi) - \alpha_1'(\xi)\} .$$

The following weak duality theorem shows that the objective of any feasible solution of the primal problem is larger than or equal to the objective of any feasible solution of the dual problem. Convexity of α_1 and concavity of α_2 are not required.

9.3.1 Theorem (Weak Duality Theorem).

$$\inf_{x \in S_1 \cap S_2} \{\alpha_1(x) - \alpha_2(x)\} \geq \sup_{\xi \in S_1' \cap S_2'} [\alpha_2'(\xi) - \alpha_1'(\xi)] .$$

Proof: Let $x_o \in S_1 \cap S_2$ and $\xi_o \in S_1' \cap S_2'$. Then we need to show that $\alpha_1(x_o) - \alpha_2(x_o) \geq \alpha_2'(\xi_o) - \alpha_1'(\xi_o)$. By definition we have $\alpha_1'(\xi_o) = \sup_{x \in S_1} \{\langle x, \xi_o \rangle - \alpha_1(x)\} \geq \langle x_o, \xi_o \rangle - \alpha_1(x_o)$ and $\alpha_2'(\xi_o) = \inf_{x \in S_2} \{\langle x, \xi_o \rangle - \alpha_2(x)\} \leq \langle x_o, \xi_o \rangle - \alpha_2(x_o)$. The last two inequalities immediately imply the desired result.

The following theorem is a strong duality theorem which shows that the dual problem possesses a solution as long as the infimum of the primal problem is finite and $\text{ri } S_1 \cap \text{ri } S_2 \neq \emptyset$. Moreover the maximum of D is equal to the infimum corresponding to the primal problem. No duality gaps are encountered.

9.3.2 Theorem (Fenchel's Duality Theorem). Let S_1 and S_2 be convex sets with $\text{ri } S_1 \cap \text{ri } S_2 \neq \emptyset$. Suppose that $\mu = \inf_{x \in S_1 \cap S_2} \{\alpha_1(x) - \alpha_2(x)\}$ is finite. Then

$$\inf_{x \in S_1 \cap S_2} \{\alpha_1(x) - \alpha_2(x)\} = \max_{\xi \in S_1' \cap S_2'} \{\alpha_2'(\xi) - \alpha_1'(\xi)\} = \alpha_2'(\xi_0) - \alpha_1'(\xi_0) \text{ for some}$$

$\xi_0 \in S_1' \cap S_2'$. If the infimum is achieved by some $x_0 \in S_1 \cap S_2$ then

$$\max_{x \in S_1} \{\langle x, \xi_0\rangle - \alpha_1(x)\} = \langle x_0, \xi_0\rangle - \alpha_1(x_0) \text{ and}$$

$$\min_{x \in S_2} \{\langle x, \xi_0\rangle - \alpha_2(x)\} = \langle x_0, \xi_0\rangle - \alpha_2(x_0) .$$

Proof: By Theorem 9.3.1, $\inf_{x \in S_1 \cap S_2} \{\alpha_1(x) - \alpha_2(x)\} \geq \sup_{x \in S_1' \cap S_2'} \{\alpha_2'(\xi) - \alpha_1'(\xi)\}$. Hence the equality in the theorem holds if we can find an $\xi_0 \in S_1' \cap S_2'$ such that $\mu = \alpha_2'(\xi_0) - \alpha_1'(\xi_0)$. Consider the function $h = \alpha_1 - \mu$ and the sets E_h and H_{α_2}. E_h and H_{α_2} have no points of their relative interiors in common because otherwise there exists an $x \in \text{ri } S_1 \cap \text{ri } S_2$ with $y > \alpha_1(x) - \mu$ and $y < \alpha_2(x)$. This however implies that $\mu > \alpha_1(x) - \alpha_2(x)$ contradicting the definition of μ. Therefore, there exists a hyperplane that properly separates E_h and H_{α_2}, i.e., there exists a nonzero (u,v) such that $\langle (u,v),(x,y)\rangle \leq 0$ for all $(x,y) \in E_h$ and $\langle (u,v),(x,y)\rangle \geq 0$ for each $(x,y) \in H_{\alpha_2}$. It is clear that $v \leq 0$ because y can be made arbitrarily large in the first inequality. We will show that v cannot be zero. If $v = 0$ then $\langle u,x\rangle \leq 0$ for each $x \in S_1$ and $\langle u,x\rangle \geq 0$ for each $x \in S_2$, i.e., $\langle u,x\rangle \leq 0$ for each $x \in S_1 - S_2$. By assumption there is an $x \in \text{ri } S_1 \cap \text{ri } S_2$ and hence $0 \in \text{ri } S_1 - \text{ri } S_2 = \text{ri}(S_1 - S_2)$. Noting that $u \neq 0$ the inequality $\langle u,x\rangle \leq 0$ for each $x \in S_1 - S_2$ shows that there is a supporting hyperplane of $S_1 - S_2$ passing through $0 \in \text{ri}(S_1 - S_2)$ which is impossible by Theorem 2.4.10. So $v < 0$. Letting $\xi_0 = u/|v|$ we have $\langle \xi_0, x\rangle - y \leq 0$ for each $(x,y) \in E_h$ and $\langle \xi_0, x\rangle - y \geq 0$ for each $(x,y) \in H_{\alpha_2}$. If we let $y = \alpha_1(x) - \mu$ in the first inequality we get $\langle \xi_0, x\rangle - \alpha_1(x) \leq -\mu$ for each $x \in S_1$ and so $\alpha_1'(\xi_0) = \sup_{x \in S_1} \{\langle \xi_0, x\rangle - \alpha_1(x)\} \leq -\mu$. This also shows that $\xi_0 \in S_1'$. Since $\langle \xi_0, x\rangle - y \geq 0$ for each $(x,y) \in H_{\alpha_2}$ we get $\langle \xi_0, x\rangle$ $\alpha_2(x) \geq 0$ for each $x \in S_2$. Therefore, $\alpha_2'(\xi_0) = \inf_{x \in S_2} \{\langle \xi_0, x\rangle - \alpha_2(x)\} \geq 0$ and

$\xi_o \in S_2'$. This shows that $\alpha_2'(\xi_o) - \alpha_1'(\xi) \geq \mu$ where $\xi_o \in S_1' \cap S_2'$. In view of Theorem 9.3.1 we conclude that $\alpha_2'(\xi_o) - \alpha_1'(\xi_o) = \mu$ and we get

$$\mu = \inf_{x \in S_1 \cap S_2} \{\alpha_1(x) - \alpha_2(x)\}$$

$$= \max_{\xi \in S_1' \cap S_2'} \{\alpha_2'(\xi) - \alpha_1'(\xi)\}$$

$$= \alpha_2'(\xi_o) - \alpha_1'(\xi_o) \quad .$$

If the inf is achieved at $x_o \in S_1 \cap S_2$ then $\mu = \alpha_1(x_o) - \alpha_2(x_o)$. This implies that $(x_o, \alpha_1(x_o) - \mu) \in E_h \cap H_{\alpha_2}$ and we conclude that $\langle (x_o, \alpha_1(x_o) - \mu), (\xi_o, -1) \rangle = 0$. This implies that $\langle \xi_o, x_o \rangle - \alpha_1(x_o) = -\mu \geq \sup_{x \in S_1} \{\langle \xi_o, x \rangle - \alpha_1(x)\}$. Since $x_o \in S_1$ then we have $\langle \xi_o, x_o \rangle - \alpha_1(x_o) = \max_{x \in S_1} \{\langle \xi_o, x \rangle - \alpha_1(x)\}$. Similarly we have $\langle \xi_o, x_o \rangle - \alpha_2(x_o) = 0 \leq \inf_{x \in S_2} \{\langle \xi_o, x \rangle - \alpha_2(x)\}$. Since $x_o \in S_2$ then $\langle \xi_o, x_o \rangle - \alpha_2(x_o) = \min_{x \in S_2} \{\langle \xi_o, x \rangle - \alpha_2(x)\}$ and the proof is complete.

9.4. Nonlinear Programming via Conjugate Functions

At this stage it is worthwhile discussing how a nonlinear programming problem can be expressed in the form:

$$\underset{x \in S_1 \cap S_2}{\text{Minimize}} \{\alpha_1(x) - \alpha_2(x)\} \; .$$

This can be done in several ways as follows. First, let $\alpha_2(x) = 0$ for each $x \in S_2$ where $S_2 = \{x : x \in X, \; g_i(x) \geq 0, \; i = 1,2,\ldots,m\}$. Also we let $S_1 = E_n$ and $\alpha_1(x) = f(x)$. Therefore, the primal problem becomes

$$\text{Minimize } f(x) \text{ subject to } x \in X, \; g_i(x) \geq 0, \quad i = 1,2,\ldots,m \quad .$$

The dual problem is to

$$\underset{\xi \in S_1' \cap S_2'}{\text{Maximize}} \{\alpha_2'(\xi) - \alpha_1'(\xi)\}$$

where

$$S_1' = \{\xi : \sup_{x \in E_n} \{\langle \xi, x \rangle - f(x)\} < \infty\}$$

$$\alpha_1'(\xi) = \sup_{x \in E_n} \{\langle \xi, x \rangle - f(x)\}$$

$$S_2' = \{\xi : \inf \{\langle \xi, x \rangle : x \in X,\ g_i(x) \geq 0,\quad i = 1,2,\ldots,m\} > -\infty\}$$

$$\alpha_2'(\xi) = \inf \{\langle \xi, x \rangle : x \in X,\ g_i(x) \geq 0,\quad i = 1,2,\ldots,m\} .$$

Note that in the above formulation, the arbitrary set X can incorporate equality constraints, nonnegativity constraints and can also handle the discrete case.

If the functions f, $-g_i$ for $i = 1,2,\ldots,m$ are convex then the evaluation of $\alpha_1'(\xi)$ is equivalent to maximizing an unconstrainted concave function, whereas the evaluation of $\alpha_2'(\xi)$ involves the minimization of a linear function over a convex set. Whether the dual problem is in fact easier or more difficult to solve than the original problem depends on the nature of the problem under consideration.

Another way to formulate a nonlinear programming problem using conjugate functions is as follows. The problem is first changed into an unconstrained problem by introducing a perturbation vector z. By taking the dual of the perturbed problem, we obtain the dual of the programming problem under consideration. Lemma 9.4.1 below gives the form of the unconstrained problem. Equivalence of the two problems is obvious.

9.4.1 Lemma. Consider problems P-1 and P-2 below

P-1: Minimize $\{f(x) : x \in X,\ g_i(x) \geq b_i,\ i = 1,2,\ldots,m\}$

P-2: Minimize $\{\alpha_1(z) - \alpha_2(z) : z \in E_m\}$

where

$$\alpha_1(z) = \inf \{f(x) : x \in X,\ g_i(x) - b_i \geq z_i,\ i = 1,2,\ldots,m\}$$

$$\alpha_2(z) = \begin{cases} 0 \text{ if } z_i \geq 0 \text{ for } i = 1,2,\ldots,m \\ -\infty \text{ otherwise} \end{cases}$$

Then Problems P-1 and P-2 are equivalent in the sense that

(i) If $\bar{x}$ solves P-1, then $\bar{z} = g(\bar{x}) = (g_1(\bar{x}),\ldots,g_m(\bar{x}))$ solves P-2

(ii) If $\bar{z}$ solves P-2 and the inf is attained at $\bar{x}$, then $\bar{x}$ solves P-1

(iii) If $\bar{x}$ solves P-1 and $\bar{z}$ solves P-2, then $f(\bar{x}) = \alpha_1(\bar{z}) - \alpha_2(\bar{z})$.

Lemma 9.4.2 below states the conjugate dual of Problem P-2 (and hence P-1).

9.4.2 Lemma. Consider the following primal problem.

P: Minimize $\{f(x) : x \in X , g_i(x) \geq b_i, i = 1,2,\ldots,m\}$ and

Then the conjugate dual problem can be stated as follows:

$$D: \text{Maximize } \{\langle u,b\rangle - \sup_x \{\langle u,g(x)\rangle - f(x) : x \geq 0\} : u \geq 0\}$$

Proof: We will first consider problem P-2 of Lemma 9.4.1, i.e., Minimize $\{\alpha_1(z) - \alpha_2(z)\}$, $z \in E_m$. We first construct α_1' and α_2' keeping in mind that $S_1 = S_2 = E_m$.

$$\begin{aligned}
\alpha_1'(u) &= \sup_z \{\langle u,z\rangle - \alpha_1(z) : z \in E_m\} \\
&= \sup_z \{\langle u,z\rangle - \inf_x \{f(x) : g(x) - b \geq z , x \in X\} : z \in E_m\} \\
&= \sup_z \{\langle u,z\rangle - \inf_{x,s} \{f(x) : g(x) - b - s = z , x \in X , s \geq 0\} : z \in E_m\} \\
&= \sup_z \{-\inf_{x,s} \{f(x) - \langle u,z\rangle : g(x) - b - s = z , x \in X , s \geq 0\} : z \in E_m\} \\
&= \sup_z \{\sup_{x,s} \{\langle u,z\rangle - f(x) : g(x) - b - s = z , x \in X , s \geq 0\} : z \in E_m\} \\
&= \sup_s \{\sup_x \{-\langle u,b\rangle + \langle u,g(x)\rangle - f(x) - \langle u,s\rangle : x \in X\} : s \geq 0\} \\
&= \sup_x \{-\langle u,b\rangle + \sup_s\{\langle u,g(x)\rangle - f(x) - \langle u,s\rangle : s \geq 0\} : x \in X\} \\
&= \begin{cases} -\langle u,b\rangle + \sup_x \{\langle u,g(x)\rangle - f(x) : x \in X\} & \text{if } u \geq 0 \\ \infty & \text{otherwise} \end{cases}
\end{aligned}$$

we now consider x_2^1.

$$\begin{aligned}
\alpha_2'(u) &= \inf_z \{\langle u,z\rangle - \alpha_2(z) : z \in E_m\} \\
&= \inf_z \{\langle u,z\rangle : z \geq 0\} \\
&= \begin{cases} 0 & \text{if } u \geq 0 \\ -\infty & \text{otherwise} \end{cases}
\end{aligned}$$

From this it is immediate that $S_1' = S_2' = E_m^+$ and hence $S_1' \cap S_2' = E_m^+$. Therefore, the dual problem can be stated as follows:

$$\underset{\xi \in S_1' \cap S_2'}{\text{maximize}} \ \alpha_2'(S) - \alpha_1'(\xi) \equiv \text{Maximize } \{\langle u,b\rangle - \sup_x\{\langle u,g(x) - f(x) : x \in X\} : u \geq 0\} .$$

The above relationship has an interesting economic interpretation due to Williams [35]. Let $X = E_n$. Consider a manufacturer who wants to produce commodities, 1,2,..., and m the demands of which are $b_1, b_2, \ldots,$ and b_m. Let the input decision vector be x, which may represent raw materials, manpower, machine hours, etc., required to do the job. Hence we have the nonnegativity constraint $x \geq 0$. Given that an input vector x is employed, then the number of units of commodity i produced is given by $g_i(x)$. Therefore, the demand constraint becomes $g(x) \geq b$. If the cost at a level x is given by $f(x)$, then the manufacturer's problem is precisely the original problem, i.e., minimize$_x$ $\{f(x) : x \geq 0, g(x) \geq b\}$. We will now discuss the interpretation of the second problem of Lemma 9.4.2. Consider a contractor who wants to rent the facilities from the manufacturer, produce the commodities, and then sell them back to him. Now the manufacturer may agree to let the contractor use his facilities if the rent paid to him by the contractor is at least equal to him maximum profit had he undertaken the whole operation by himself. In other words, suppose that the contractor quotes a price $\xi_i \geq 0$ for commodity i, then had he produced the commodities, the manufacturer could have achieved a maximum profit which is equal to sup $\{\langle \xi, g(x) \rangle - f(x) : x \geq 0\}$. This profit is exactly the amount of rent the manufacturer should ask for. Therefore, from the contractor's point of view, the problem is reduced to quoting the optimal price vector ξ which maximizes his profit, namely the income $\langle \xi, b \rangle$ minus the rent. Therefore, the contractor's problem becomes to maximize $\{\langle \xi, b \rangle - \sup \{\langle \xi, g(x) \rangle - f(x) : x \geq 0\} : \xi \geq 0\}$, or precisely problem D above. Hence the weak duality theorem 9.3.1 states that the manufacturer's minimum cost is larger than or equal to the contractor's maximum profit. Moreover, if the functions involved satisfy the hypotheses of Theorem 9.3.2, then the two optimal solutions are equal.

SELECTED REFERENCES

[References marked with an asterisk contain more detailed and exhaustive references on convex analyses and extremization theory.]

[1] Abadie, J., "On the Kuhn Tucker Theorem" in Nonlinear Programming, J. Abadie (ed.), North-Holland Publishing Co., Amsterdam, (1967).

[2] Arrow, K. J., Hurwicz, L. and Uzawa, H., "Constraint Qualifications in Maximization Problems," Naval Research Logistics Quarterly, Vol. 8 (1961), pp. 175-191.

[3] ―――――― and Uzawa, H., "Constraint Qualifications in Maximization Problems, II," Technical Report No. 84, Institute for Mathematical Studies in Social Sciences, Stanford (1960).

[4] Bazaraa, M. S., Goode, J. J., and Shetty, C. M., "Optimality Criteria Without Differentiability," Operations Research, Vol. 19 (1971), pp. 77-86.

[5] Bazaraa, M. S., Goode, J. J., and Shetty, C. M., "Constraint Qualifications Revisited," Management Science, Vol. 18 (1972), pp. 567-573.

*[6] Berman, A., Cones, Matrices and Mathematical Programming, Lecture Notes in Economics and Mathematical Systems, No. 79, Springer-Verlag, (1973).

*[7] Canon, M. D., Cullum, Jr., C. D. and Polak, E., Theory of Optimal Control and Mathematical Programming, McGraw-Hill Book Co., (1970).

[8] Cottle, R. W., "A Theorem of Fritz John in Mathematical Programming," RAND Corporation Memorandum RM-3858-PR, (1963).

[9] Cullen, C., Matrices and Linear Transformations, Addison-Wesley Publishing Company (1966).

[10] Evans, J. P., "On Constraint Qualifications in Nonlinear Programming," Center for Mathematical Studies in Business and Economics, University of Chicago (1969).

[11] Everett, H., "Generalized Lagrange Multiplier Method for Solving Problems of Optimum Allocation of Resources," Operations Research, 11 (1963), pp. 399-417.

[12] Falk, J. E., "Lagrange Multipliers and Nonlinear Programming," Math. Anal. and Appl., Vol. 19, (1967), pp. 141-159.

[13] Falk, J. E., "Lagrange Multipliers and Nonconvex Programs," SIAM J. Control, Vol. 7, (1969), pp. 534-545.

[14] Fenchel, W., Convex Cones, Sets and Functions, Lecture Notes, Princeton University (1951).

*[15] Fiacco, A. V. and McCormick, G. P., Nonlinear Programming: Sequential Unconstrained Minimization Techniques, John Wiley, (1968).

*[16] Geoffrion, A. M., "Duality in Nonlinear Programming: A Simplified Applications-Oriented Development," SIAM Review, Vol. 13, No. 1, (1971).

*[17] Girsanov, I. V., Lectures on Mathematical Theory of Extremum Problems, Lecture Notes in Economics and Mathematical Systems, No. 67, Springer-Verlag, New York (1972).

[18] Gould, F. J., and Tolle, J. W., "A Necessary and Sufficient Qualification for Constrained Optimization," SIAM J. Appl. Math., 20 (1971), pp. 164-172.

[19] Guignard, M., "Generalized Kuhn-Tucker Conditions for Mathematical Programming Problems in a Banach Space," SIAM Journal of Control, Vol. 7 (1969), pp. 232-241.

*[20] Hadley, G., Nonlinear and Dynamic Programming, Addison-Wesley, (1964).

[21] John, F., "Extremum Problems with Inequalities as Side Conditions," in Studies and Essays, Courants Anniversary Volume, Friedricks, Neugebauer, Stoker (eds.), Wiley, (1948).

[22] Karlin, S., Mathematical Methods and Theory in Games, Programming and Economics, Vol. I, Addison-Wesley, Reading, Mass, (1959).

[23] Kuhn, H. W. and Tucker, A. W., "Nonlinear Programming," in Proceedings of the Second Berkeley Symposium on Mathematical Statistics and Probability, University of California Press, Berkeley (1951), pp. 481-492.

*[24] Luenberger, D. G., Optimization by Vector Space Methods, John Wiley, New York (1969).

*[25] Mangasarian, O. L., Nonlinear Programming, McGraw-Hill Book Company, Inc., (1969).

[26] ―――――――― and Fromovitz, S., "The Fritz John Necessary Optimality Conditions in the Presence of Equality and Inequality Constraints," J. Math. Anal. Appl., Vol. 17 (1967), pp. 37-47.

*[27] Nikaido, Convex Structures and Economic Theory, Academic Press (1968).

[28] Peterson, D. W., "A Review of Constraint Qualifications in Finite-Dimensional Spaces," SIAM REVIEW, Vol. 15, No. 3 (1973).

*[29] Rockafellar, R. T., Convex Analysis, Princeton University Press, (1970).

[30] Rockafellar, R. T., "Duality in Nonlinear Programming," Maths of the Decision Sciences, Part I, G. B. Dantzig and A. F. Veinott (eds.), American Math. Society, Proudence, (1968), pp. 401-422.

*[31] Roode, J. D., "Generalized Lagrangian Functions in Mathematical Programming," Thesis, Leiden, (1968).

[32] Slater, M., Lagrange Multipliers Revisited: A Contribution to Nonlinear Programming, Cowles Commission Discussion Paper, Mathematics 403, (1950).

*[33] Stoer, J., and Witzgall, C., Convexity and Optimization in Finite Dimensions I, Springer-Verlag, New York, (1970).

[34] Uzawa, H., The Kuhn-Tucker Theorem in Concave Programming, in Studies in Linear and Nonlinear Programming, Arrow, Hurwicz and Uzawa (eds.), Stanford University Press, (1958).

[35] Williams, A. C., "Nonlinear Activity Analysis and Duality," Proceedings of the Sixth International Symposium on Math. Programming, Princeton, (1967).

*[36] Zangwill, W. I., Nonlinear Programming: A Unified Approach, Prentice-Hall, Inc., Englewood Cliffs, N. J., (1969).

*[37] Zoutendijk, G., Methods of Feasible Directions: A Study in Linear and Nonlinear Programming, Elserier Publishing Co., Amsterdam, (1960).

Vol. 59: J. A. Hanson, Growth in Open Economies. V, 128 pages. 1971.

Vol. 60: H. Hauptmann, Schätz- und Kontrolltheorie in stetigen dynamischen Wirtschaftsmodellen. V, 104 Seiten. 1971.

Vol. 61: K. H. F. Meyer, Wartesysteme mit variabler Bearbeitungsrate. VII, 314 Seiten. 1971.

Vol. 62: W. Krelle u. G. Gabisch unter Mitarbeit von J. Burgermeister, Wachstumstheorie. VII, 223 Seiten. 1972.

Vol. 63: J. Kohlas, Monte Carlo Simulation im Operations Research. VI, 162 Seiten. 1972.

Vol. 64: P. Gessner u. K. Spremann, Optimierung in Funktionenräumen. IV, 120 Seiten. 1972.

Vol. 65: W. Everling, Exercises in Computer Systems Analysis. VIII, 184 pages. 1972.

Vol. 66: F. Bauer, P. Garabedian and D. Korn, Supercritical Wing Sections. V, 211 pages. 1972.

Vol. 67: I. V. Girsanov, Lectures on Mathematical Theory of Extremum Problems. V, 136 pages. 1972.

Vol. 68: J. Loeckx, Computability and Decidability. An Introduction for Students of Computer Science. VI, 76 pages. 1972.

Vol. 69: S. Ashour, Sequencing Theory. V, 133 pages. 1972.

Vol. 70: J. P. Brown, The Economic Effects of Floods. Investigations of a Stochastic Model of Rational Investment. Behavior in the Face of Floods. V, 87 pages. 1972.

Vol. 71: R. Henn und O. Opitz, Konsum- und Produktionstheorie II. V, 134 Seiten. 1972.

Vol. 72: T. P. Bagchi and J. G. C. Templeton, Numerical Methods in Markov Chains and Bulk Queues. XI, 89 pages. 1972.

Vol. 73: H. Kiendl, Suboptimale Regler mit abschnittweise linearer Struktur. VI, 146 Seiten. 1972.

Vol. 74: F. Pokropp, Aggregation von Produktionsfunktionen. VI, 107 Seiten. 1972.

Vol. 75: GI-Gesellschaft für Informatik e.V. Bericht Nr. 3. 1. Fachtagung über Programmiersprachen · München, 9.–11. März 1971. Herausgegeben im Auftrag der Gesellschaft für Informatik von H. Langmaack und M. Paul. VII, 280 Seiten. 1972.

Vol. 76: G. Fandel, Optimale Entscheidung bei mehrfacher Zielsetzung. II, 121 Seiten. 1972.

Vol. 77: A. Auslender, Problèmes de Minimax via l'Analyse Convexe et les Inégalités Variationelles: Théorie et Algorithmes. VII, 132 pages. 1972.

Vol. 78: GI-Gesellschaft für Informatik e.V. 2. Jahrestagung, Karlsruhe, 2.–4. Oktober 1972. Herausgegeben im Auftrag der Gesellschaft für Informatik von P. Deussen. XI, 576 Seiten. 1973.

Vol. 79: A. Berman, Cones, Matrices and Mathematical Programming. V, 96 pages. 1973.

Vol. 80: International Seminar on Trends in Mathematical Modelling, Venice, 13–18 December 1971. Edited by N. Hawkes. VI, 288 pages. 1973.

Vol. 81: Advanced Course on Software Engineering. Edited by F. L. Bauer. XII, 545 pages. 1973.

Vol. 82: R. Saeks, Resolution Space, Operators and Systems. X, 267 pages. 1973.

Vol. 83: NTG/GI-Gesellschaft für Informatik, Nachrichtentechnische Gesellschaft. Fachtagung „Cognitive Verfahren und Systeme", Hamburg, 11.–13. April 1973. Herausgegeben im Auftrag der NTG/GI von Th. Einsele, W. Giloi und H.-H. Nagel. VIII, 373 Seiten. 1973.

Vol. 84: A. V. Balakrishnan, Stochastic Differential Systems I. Filtering and Control. A Function Space Approach. V, 252 pages. 1973.

Vol. 85: T. Page, Economics of Involuntary Transfers: A Unified Approach to Pollution and Congestion Externalities. XI, 159 pages. 1973.

Vol. 86: Symposium on the Theory of Scheduling and its Applications. Edited by S. E. Elmaghraby. VIII, 437 pages. 1973.

Vol. 87: G. F. Newell, Approximate Stochastic Behavior of n-Server Service Systems with Large n. VII, 118 pages. 1973.

Vol. 88: H. Steckhan, Güterströme in Netzen. VII, 134 Seiten. 1973.

Vol. 89: J. P. Wallace and A. Sherret, Estimation of Product. Attributes and Their Importances. V, 94 pages. 1973.

Vol. 90: J.-F. Richard, Posterior and Predictive Densities for Simultaneous Equation Models. VI, 226 pages. 1973.

Vol. 91: Th. Marschak and R. Selten, General Equilibrium with Price-Making Firms. XI, 246 pages. 1974.

Vol. 92: E. Dierker, Topological Methods in Walrasian Economics. IV, 130 pages. 1974.

Vol. 93: 4th IFAC/IFIP International Conference on Digital Computer Applications to Process Control, Part I. Zürich/Switzerland, March 19–22, 1974. Edited by M. Mansour and W. Schaufelberger. XVIII, 544 pages. 1974.

Vol. 94: 4th IFAC/IFIP International Conference on Digital Computer Applications to Process Control, Part II. Zürich/Switzerland, March 19–22, 1974. Edited by M. Mansour and W. Schaufelberger. XVIII, 546 pages. 1974.

Vol. 95: M. Zeleny, Linear Multiobjective Programming. X, 220 pages. 1974.

Vol. 96: O. Moeschlin, Zur Theorie von Neumannscher Wachstumsmodelle. XI, 115 Seiten. 1974.

Vol. 97: G. Schmidt, Über die Stabilität des einfachen Bedienungskanals. VII, 147 Seiten. 1974.

Vol. 98: Mathematical Methods in Queueing Theory. Proceedings 1973. Edited by A. B. Clarke. VII, 374 pages. 1974.

Vol. 99: Production Theory. Edited by W. Eichhorn, R. Henn, O. Opitz, and R. W. Shephard. VIII, 386 pages. 1974.

Vol. 100: B. S. Duran and P. L. Odell, Cluster Analysis. A Survey. VI, 137 pages. 1974.

Vol. 101: W. M. Wonham, Linear Multivariable Control. A Geometric Approach. X, 344 pages. 1974.

Vol. 102: Analyse Convexe et Ses Applications. Comptes Rendus, Janvier 1974. Edited by J.-P. Aubin. IV, 244 pages. 1974.

Vol. 103: D. E. Boyce, A. Farhi, R. Weischedel, Optimal Subset Selection. Multiple Regression, Interdependence and Optimal Network Algorithms. XIII, 187 pages. 1974.

Vol. 104: S. Fujino, A Neo-Keynesian Theory of Inflation and Economic Growth. V, 96 pages. 1974.

Vol. 105: Optimal Control Theory and its Applications. Part I. Proceedings 1973. Edited by B. J. Kirby. VI, 425 pages. 1974.

Vol. 106: Optimal Control Theory and its Applications. Part II. Proceedings 1973. Edited by B. J. Kirby. VI, 403 pages. 1974.

Vol. 107: Control Theory, Numerical Methods and Computer Systems Modeling. International Symposium, Rocquencourt, June 17–21, 1974. Edited by A. Bensoussan and J. L. Lions. VIII, 757 pages. 1975.

Vol. 108: F. Bauer et al., Supercritical Wing Sections II. A Handbook. V, 296 pages. 1975.

Vol. 109: R. von Randow, Introduction to the Theory of Matroids. IX, 102 pages. 1975.

Vol. 110: C. Striebel, Optimal Control of Discrete Time Stochastic Systems. III. 208 pages. 1975.

Vol. 111: Variable Structure Systems with Application to Economics and Biology. Proceedings 1974. Edited by A. Ruberti and R. R. Mohler. VI, 321 pages. 1975.

Vol. 112: J. Wilhlem, Objectives and Multi-Objective Decision Making Under Uncertainty. IV, 111 pages. 1975.

Vol. 113: G. A. Aschinger, Stabilitätsaussagen über Klassen von Matrizen mit verschwindenden Zeilensummen. V, 102 Seiten. 1975.

Vol. 114: G. Uebe, Produktionstheorie. XVII, 301 Seiten. 1976.

Vol. 115: Anderson et al., Foundations of System Theory: Finitary and Infinitary Conditions. VII, 93 pages. 1976

Vol. 116: K. Miyazawa, Input-Output Analysis and the Structure of Income Distribution. IX, 135 pages. 1976.

Vol. 117: Optimization and Operations Research. Proceedings 1975. Edited by W. Oettli and K. Ritter. IV, 316 pages. 1976.

Vol. 118: Traffic Equilibrium Methods, Proceedings 1974. Edited by M. A. Florian. XXIII, 432 pages. 1976.

Vol. 119: Inflation in Small Countries. Proceedings 1974. Edited by H. Frisch. VI, 356 pages. 1976.

Vol. 120: G. Hasenkamp, Specification and Estimation of Multiple-Output Production Functions. VII, 151 pages. 1976.

Vol. 121: J. W. Cohen, On Regenerative Processes in Queueing Theory. IX, 93 pages. 1976.

Vol. 122: M. S. Bazaraa, and C. M. Shetty, Foundations of Optimization VI. 193 pages. 1976